Healthy Soil

THE BEST OF
FINE GARDENING

Healthy Soil

The Taunton Press

Cover photo: Susan Kahn

Back-cover photos: top, Staff;
bottom, Susan Kahn;

Illustration: Michael Rothman

for fellow enthusiasts

First printing: 1995
Printed in the United States of America

A FINE GARDENING Book

FINE GARDENING® is a trademark of The Taunton Press, Inc.,
registered in the U.S. Patent and Trademark Office.

The Taunton Press
63 South Main Street
Box 5506
Newtown, CT 06470-5506

Library of Congress Cataloging-in-Publication Data

Healthy soil.
    p.   cm. — (The Best of Fine gardening)
    Articles originally appeared in Fine gardening magazine.
    "A Fine gardening book"— T.p. verso.
    Includes index.
    ISBN 1-56158-101-1
    1. Garden soils. 2. Soil Management. 3. Gardening.
    I. Fine gardening. II. Series.
S596.75.H43   1995   95-11642
635'.044 — dc20   CIP

Contents

Introduction 7

Organic Matters 8
Build your soil and feed your plants

The Colorful Soil 10
Darker isn't always better

The World Under Our Feet 13
Gardeners can encourage hidden allies in the soil

Home Soil Testing 16
Are kits worth the effort?

Evaluating Potting Soil 21
What makes a good mix?

Soil Drainage 26
What's soggy, what's not, and what to do about it

Composting Leaves 29
Recycling fall's bounty to improve your soil

Get Started in Composting 32
Turn yard waste into a valuable soil booster

Making Hot Compost 36
Transform yard wastes into a valuable soil amendment

Cover Crops 40
Using plants to build the soil

Microscopic Partnership 45
Fungi that help plants grow better

Soil Amendments 48
How lime, sulfur and organic matter improve poor soil

A Buyer's Guide to Fertilizers 52

Slow-Release Fertilizers 56
One application nourishes plants for months

Fertilizing Trees Makes a Difference 58
A little goes a long way

Seaweed Comes Ashore 61
A natural soil amendment and plant growth stimulant

A Mulch Primer 64
Year-round cover aids plants and soil

Mulch, Don't Dig 68
Layers of newspaper prepare a new garden bed

Peat Moss 71
Does this widely available soil amendment have uses in your garden?

No-Till Gardening 74
Soil improvement from the top down

Gardening amid Tree Roots 76
Understand the risks before you dig, water or apply herbicides

Managing Water in Arid Gardens 80
Simple, efficient, economical methods

Reclaiming a Lifeless Soil 85
A summer's work makes a difference

No More Lawn-Mower Bag 89
Grass clippings nourish the lawn and don't cause thatch

Houseplant Hydroponics 91
Soilless system in a pot

Index 94

Introduction

Here are the best articles on working with soil presented by *Fine Gardening* magazine in its first six years of publication.

In this beautifully illustrated collection, expert home gardeners, horticulturists and landscape contractors provide answers to many common questions people have about the fundamentals of soil structure and composition. But the bulk of this book deals with soil improvement through the use of compost and other amendments, as well as specific cultural practices. Ranging from the tried-and-true to the innovative, many of the approaches recommended by the authors are sure to suit your situation.

You'll find the articles in this collection especially helpful and inspiring because they are the work of enthusiasts, gardeners who have given much thought to the hows and whys of good horticulture. Sharing their hard-won experience, the authors tell you how to provide the best possible soil for your own plants.

The editors of *Fine Gardening* hope you'll experiment with the ideas in this collection of articles. No matter which you choose to try, your efforts will be rewarded.

"The Best of Fine Gardening" series collects articles from back issues of *Fine Gardening* magazine. A note on p. 96 gives the date of first publication for each article; product availability, suppliers' addresses and prices may have changed since original publication. This book is the seventh in the series. The next title is "The Best of Fine Gardening on Garden Tools & Equipment."

Organic Matters
Build your soil and feed your plants

by Robert Parnes

Looking at an abundant garden filled with vigorous, leafy, blooming plants, it's hard to imagine that an equal but invisible amount of life teems in the soil below. We don't see the microorganisms—fungi, bacteria and so on—that are active in a good garden soil, but their presence is valuable in many ways. Most important, they build a good soil structure, one that allows plant roots to grow unhindered, and permits the flow of water and air.

To encourage these soil organisms, we need to provide them with an energy source—specifically, organic residues rich in carbon, such as hay, manure or compost. These residues also provide nutrients for plants. With some selectivity, you can supply enough organic residues to meet the energy needs of the soil microorganisms, and at the same time supply enough nutrients to meet the fertilizer requirements of your crop plants (although some soils will still need lime). In the process of feeding your soil, you can also fertilize your plants.

What is food for the soil?

Food provides both energy and nutrients. It's important to distinguish between them. To draw an analogy to our own diet, energy is the caloric content of food, and nutrients are like the vitamins and minerals. Energy is needed for metabolism and growth. Plants get energy from the sun, and store it in the form of carbohydrates. By consuming these carbohydrates, soil organisms recover the energy that the plants have stored.

Nutrients are nitrogen and minerals such as phosphorus, potassium, sulfur, magnesium and calcium. Although the amounts required are small, these nutrients play important roles in living organisms. For example, nitrogen is built into proteins, potassium affects the balance of water pressure inside and outside of cells, and magnesium is the key ingredient in chlorophyll.

Soil microorganisms can get enough nutrients from the soil, but we have to supply them with energy. In the vast majority of garden soils, lack of carbohydrates limits biological activity. If we add organic residues containing carbohydrates, biological activity increases substantially, until the energy in the carbohydrates is used up.

One way to assess the energy potential of organic residues is by looking at the ratio

---

Common organic residues can supply almost all the nutrients a plant needs.

---

of carbon to nitrogen (the C/N ratio) present in the residue. The higher the ratio, the more energy the material contains, and the longer it will take microorganisms to digest it. The lower the ratio, the less energy the material contains, and the faster its nitrogen will be released for the plants to use. The C/N ratio varies widely, from as low as 3 for a nitrogen-rich product like blood meal, to as high as 200 for carbon-rich wood chips. Fresh hay, fresh manure, and compost have C/N ratios in the range of about 15 to 30, and balance the energy needs of the soil organisms with the nitrogen needs of the plants.

Using organic residues to fertilize your plants

Average vegetable crops in an average garden need an annual supply of about 80 lb. per acre of nitrogen, 30 lb. per acre of phosphorus and 120 lb. per acre of potassium or potash. Some nutrients are present in the soil of any garden, but they are continually being lost by harvest, leaching, erosion and other means. To sustain an intensive garden, at least as many nutrients must be replaced as are taken out. The chart on the facing page lists the average nitrogen, phosphate and potash content of several organic residues. Three of the best are fresh hay, animal manure and compost.

An annual dose of organic residues is as valuable in the perennial border as it is in the vegetable patch. If you're starting a new garden from scratch, whether it's for vegetables or flowers, begin by adding large amounts of organic residues to supply energy for the microorganisms while their activities improve the structure of the soil. Soil microorganisms can consume remarkable quantities of hay, manure or other matter in a season—a bulky layer of energy-rich residues ends up as a thin film of nutrient-laden humus. For the first year or two, adding residues may not supply enough nutrients in time to meet your plants' needs, and you may choose to use a small amount of supplemental fertilizer. After a garden has been developed, however, the amounts of residues recommended below will maintain the soil in good condition and supply enough nutrients for the plants. The suggested rates of annual application apply to soils in the Midwest, Northeast and parts of the Northwest. Soils in the South and Far West require more of everything—twice as much wouldn't hurt.

Hay—Fresh hay is underappreciated as a fertilizer. It offers a complete range of nutrients without the mess of handling manure or the labor of composting. Hay supplies a maximum amount of energy to the soil microorganisms, and since it decomposes rapidly once it gets onto the ground, it can also release a significant amount of nitrogen for the plants. If it has a good green color, hay is a good source of both nutrients and energy. Aged, discolored, moldy or rotten hay is acceptable, but probably has lost some nutrients.

Because of its coarse texture, hay is difficult to work into the soil. It's more easily used as a mulch. It doesn't require tilling, and you can leave it on the ground all year long. In the spring, it's a good idea to pull back the mulch and give the uncovered soil time to warm up before planting hot-weather crops. In the summer, use the mulch to keep the soil cool and moist. One drawback of a mulch is that pests are attracted to its cool dampness; where slugs are a menace ordinarily, they can be a nightmare under hay. Hay is likely to produce weed seedlings, although continual smothering by additional mulch should control them. Also, hay is bulkier than manure or compost and requires more trips to haul it to the garden.

An average bale of hay is roughly 14 in. by 20 in. by 32 in. and may weigh 40 lb. Sixteen bales are enough to spread a 1-in. layer of hay as a mulch on 1,000 sq. ft. of garden. (A bale will cleave nicely into wafers of compressed hay of whatever thickness you desire.) In an established gar-

COMMONLY AVAILABLE ORGANIC RESIDUES	Nutrient content in lb./ton of:			C/N ratio	Lb./1,000 sq. ft. needed to meet plant requirements per crop		
	Nitrogen	Phosphate	Potash		Nitrogen*	Phosphate	Potash
Animal manure							
Cow	11	3	9	18	340	470	620
Horse	12	4	10	22	320	360	560
Pig	13	8	7	14	290	180	800
Poultry	30	20	10	7	240	70	560
Sheep	20	5	20	16	190	280	280
Compost	24	8	60	17	160	180	90
Fresh hay							
Grass	25	11	38	32	150	130	150
Legume	50	13	33	16	70	110	170
Leaves							
Deciduous	16	3	10	65	N/A	470	560
Evergreen	16	3	3	65	N/A	470	1900
Straw	13	5	26	72	N/A	280	220

As much as half of the nitrogen in manure can be lost between the animal and the plant, so the quantity needed to meet the nitrogen requirement with manure may need to be doubled.

den, that should supply, according to the chart and with no losses, about four times the nutrients required by the average vegetable crop. Adding more hay wouldn't hurt and would more effectively smother weeds. Some gardeners successfully mulch with full bales. Unbaled hay is much looser than baled hay, so I recommend applying it in a layer 4 in. to 6 in. deep.

Manure—In addition to the major nutrients, animal manure includes numerous organic enzymes and a great many active microorganisms. Fresh manure contains significant amounts of nitrogen, but much of it is lost before becoming available to plants. Even if half the nitrogen is lost, however, most manure supplies about the right proportions of nutrients.

The major drawback to fresh manure is its odor and stickiness. Probably the best method for using it is to spread it on the soil and turn it under immediately, preferably in the fall, after harvesting your vegetables or cleaning up your flower border. Typical rates for applying fresh manure to supply nutrients are 500 lb. to 1,000 lb. per 1,000 sq. ft. for non-poultry manure, and roughly one-fifth that amount for poultry manure. (A cubic yard of fresh horse manure weighs approximately 800 lb.; of other manures, about 1600 lb.) If you spread manure at the end of the growing season, it's a good idea to sow a cover crop such as vetch or rye, which will take up and store nutrients through the winter.

You can also apply manure a week or two before planting in the spring, but hold the rates of application down to 500 lb. per 1,000 sq. ft. of non-poultry or 100 lb. per 1,000 sq. ft. of poultry manure.

Compost—Composting is ideal for turning unsuitable residues such as kitchen wastes and tough, stalky crop residues into a pleasant, earthy substance rich in energy, nutrients and biological activity—indeed a magic process, if done correctly. Hot composting, in which a high initial level of biological activity raises the temperature of the decomposing materials to about 150°F, kills weed seeds and pests, but it's also likely to result in large losses of nitrogen. Hot composting requires a lot of planning, attention and work, and is overwhelming for most people trying to supply a medium-size garden. I prefer cold composting, because it's less work—I simply make piles. The process of cold composting may take as long as a year, but it may retain more nitrogen than hot compost does.

Typical rates for applying compost range from 500 lb. to 1,000 lb. per 1,000 sq. ft. (A cubic yard of compost weighs about 800 lb.) A 1,000-lb. application will supply roughly six times the nutrients required for the average vegetable crop, but in general there's no risk of overdoing it with compost. It can be applied as a mulch or worked into the soil at any time of the year.

Other organic residues—Lawn clippings, straw, tree leaves and wood products can all provide energy for soil organisms. Dried lawn clippings, like grassy hay, also have fertilizer value. Thick layers of green clippings pack into dense water-repellent masses, but dried clippings can be used as a mulch. Straw (the dried stems and leaves of crops harvested for seed or grain) supplies less nitrogen and phosphorus than hay does, but it's more likely than hay to be free of weed seeds and it makes a good mulch. Tree leaves contain copper, iron, manganese and other trace minerals. Shredding them helps prevent matting. They can be spread as a mulch, mixed into the soil or composted. Fresh wood chips and sawdust make good mulches and provide energy to the soil, but don't contain enough nutrients to qualify as fertilizers.

There can be a problem with adding large amounts of carbon-rich residues, such as straw, leaves and wood chips, to the soil. In the process of digesting the residues, the soil organisms will find all the nutrients they need to grow and multiply, and meanwhile the plants may be deprived. Plants simply can't compete with microorganisms when it comes to nutrient uptake—the microorganisms always take what they need, and the plants get what's left over.

In particular, nitrogen, which would otherwise be available to plants, may be secured by the soil organisms for a period of weeks or months. Eventually the microbial activity will subside and the nitrogen will be released for the plants again. Some gardeners add extra nitrogen to compensate for this temporary tie-up, timing the application to meet the needs of the growing plants.

Other nutrient needs

In years of doing soil tests at Woods End Laboratory in Maine, I observed that a soil treated regularly with a variety of organic residues will almost always have enough potassium and trace elements, and rarely will be low in phosphorus, though I must hedge a bit here, as some gardeners automatically supplement organic residues with rock phosphate or bone meal. But in usual practice, the only nutrients that may be low in soils well treated with organic residues are nitrogen and, in much of the eastern half of North America, magnesium. You can use a small amount of fertilizer to provide the nitrogen. The magnesium can be supplied by a high-magnesium limestone.

Consequently, the use of fertilizer becomes a secondary question. What's most important is adding enough organic residues to supply energy to the soil. Without them, no fertilizer of any amount will prevent the deterioration of the soil, and with them at most a minimal amount of fertilizer is necessary. □

Robert Parnes lives in Mechanic Falls, Maine. His book, Organic & Inorganic Fertilizers, *costs $15 postpaid from Woods End Agricultural Institute, RFD 1, Box 4050, Mt. Vernon, ME 04352.*

The Colorful Soil
Darker isn't always better

by Francis D. Hole

For about 45 years, I've used the language of research and teaching, along with song and drama, to celebrate the soil and help others understand and appreciate it as much as I do. Soil-watching is a great pleasure. I enjoy the beauty of the soil for its own sake. Yet my experiences with soil have been enriched as I've learned more about how it's formed and how to interpret what I see. Even gardeners and farmers, who have intimate contact with the surface soil, are often unfamiliar with much of the vital kingdom beneath our feet.

Soil supports life on land, enduring wherever it's well protected by a covering of vegetation, natural mulch or litter. The soil nourishes plants; plants protect the soil. This cooperative enterprise flourishes in well-managed gardens and farms, as well as in forests and grasslands. Through food crops, we draw on the soil for our nutrition. The soil is, both literally and figuratively, the root domain.

It's taken millions of years for enormous quantities of solid, seemingly unyielding rock to be transformed into particles of soil. To imagine the process of soil-making, visualize a cubic foot of granite. Nearly all of the 175 lb. of tightly interlocked mineral grains are isolated from the environment—some three million sand-size grains waiting to be set free by the slow processes of disintegration. As sun, wind, water, roots and organic acids work on the rock, these grains separate, breaking down further into finer sand, silt and clay particles. Sand particles are the coarsest, being 2mm to 0.05mm, or $\frac{1}{12}$ in. to $\frac{1}{500}$ in., in diameter. Silt is comprised of medium-fine particles, 0.05mm to 0.002mm in diameter, and clay is very fine, with particles less than 0.002mm in diameter.

When these millions of particles are mixed with organic materials, we have soil. Decomposed vegetable and animal matter, or humus, gives soils their black, dark-gray and dark-brown colors. The

This black muck soil took thousands of years to form from plant remains that accumulated underwater in a wetland near Belleglade, Florida. Crops can grow on soils like these after the soil has been artificially drained, but this process exposes the organic materials to air, accelerates their decomposition and decreases long-term productivity.

mineral particles and organic matter usually cluster together in pea- or block-shaped units that are fashioned by earthworms, moles, insects and roots, and by the swelling and shrinking of the soil. The spaces between these clusters of particles double the volume of the granite, making room for water, air, roots and animals, which the rocks, in their density, would not accept. With an increased volume and surface area, the soil acts like a sponge, absorbing water and storing nutrients that were released from the minerals.

As we travel through the landscape, the most striking aspect of soil is its color. We may be surprised that soils in different landscapes can have different colors. Foundations of houses are splashed with red soil in Alabama, black in Minnesota and yellow in parts of North Carolina.

Soil color frequently has a special meaning for gardeners and farmers, who use it as an index of the fertility of their land, often praising black soil for its ex-

ceptional productivity. They add animal manure, compost and cover crops, hoping they can transform light-colored soils into black ones. Somehow, an unspoken consensus has developed among growers in many parts of the world that the best soil is black.

Yet it's not quite that simple. Each of the four common soil-color groups—black, white, red and yellow, and pale bluish-gray—has a unique origin. No matter what you add to some soils, they won't turn black. Depending on what you want to grow, red or white soil might be as good or better. Black soil is sometimes fertile, but not always. It depends on how the soil was formed, what's been growing there and how the land has been managed. Soils of the same color in different locations may be very different. For example, a young black chernozem soil in the Russian Ukraine is naturally fertile; nutrients haven't yet been leached out. On the other hand, a look-alike old black soil in southern Brazil is infertile; thousands of years of leaching have left it very acid and low in plant nutrients.

Black soils—Black soils are colored by carbon-rich humus, the complex residue from the decomposition of plant materials. The humus can stain the surfaces of light-colored minerals, such as quartz. It can also exist on its own, as in the peat and muck soils of wetlands.

In some soils, the dark color comes both from surface staining of mineral particles and from vast stores of organic matter. For thousands of years, prairie grasses and flowering plants have covered the landscape near the village of Black Earth, Wisconsin, depositing humus to a depth of several feet. This soil is naturally fertile—crops of oats, hay and corn have been grown here ever since the European settlers switched a century ago from cultivating wheat fields to establishing dairy farms. But the immediate fertility is not due to the humus, despite its high level of nutrient reserves and black color. The quick fertility that supports crop growth is due to the ample supply of mineral particles, which attract and release plant nutrients more rapidly than the humus does. (The slower release of nutrients from the humus and clay in this soil is, of course, beneficial in the long term.)

Another group of black to dark-brown soils, peats and mucks, are usually formed in bogs and marshes. In these waterlogged habitats, partially decomposed vegetation accumulates. In the absence of oxygen, it rots slowly and builds up a wet organic mass called peat. (Muck is just the surface layers of more decomposed peat).

It may be surprising that these bog and marsh soils, referred to as organic soils (containing 30% or more organic matter by dry weight), are naturally infertile. Their nutrients are locked up in tissues that decompose too slowly to release enough to satisfy crop needs in any given year. But when these soils are drained and fertilizer is added, they're very productive for growing carrots and specialty crops such as mint and cranberries. Roots easily penetrate these unbelievably soft, spongy soils, and it's easy to control the water supply with ditches. Once these soils are drained, however, oxygen enters them, the organic matter decomposes more quickly and the whole spongy soil mass "evaporates" in a few decades.

White soils—If we generally think of soil as dirty-colored, then we're startled to see white soil, such as that found in White Earth, Minnesota, a village named for the white soil in the nearby coniferous-hardwood forests. Pioneers wondered if the white, ashlike layer buried just beneath the leaf and needle litter on the forest floor and extending several inches down was ash from past forest fires. (Even the official soil-classification name for this

The thick white layer of soil in a sandy Florida forest was formed from white quartz minerals that originally came from decomposing sandstone and granite rock. The middle layer of sand was stained black by humus, which washed down from the soil surface. Below this, a coating of iron oxide gives the soil a yellowish-brown color.

type of white soil, podzol, reflects its appearance. Coined by Russian peasants long ago, this word means "ash under.") But a century of soil studies has shown that bits of charcoal, not ash, are the most enduring contribution of a forest fire to the soil.

The pale soil is, in fact, a bed of clean, glistening quartz sand that looks somewhat like white crushed glass. The White Earth soil initially was colored by the quartz, a mineral that comes from decomposing sandstone and granite, but undoubtedly it has undergone several color changes. The original white earth, formed thousands of years ago under sedges, grasses and deciduous trees, gradually became yellowish-brown, stained by a delicate coating of iron released from dark minerals in the granite. This film of iron oxide slowly built up on the surface of the many small rock particles in the upper few inches of the soil.

Once the pine forest invaded the landscape thousands of years ago, rainwater soaked down through the rotting pine

needles and birch leaves, dissolving complex organic compounds in them. These weak acids broke down the iron oxide film, removing the yellowish-brown coating from the sand grains much the same way that lemon juice can wash rust stains out of a white handkerchief. Then the soil returned to its original white color.

This is a mineral soil (as distinguished from an organic soil), composed of 70% to 99% particles of rocks and minerals and only 1% to 30% organic matter. Some mineral soils have high nutrient levels, but the White Earth forest soils are nutrient-poor—pine and birch roots don't add much organic matter on the soil surface, and not much of the leaf litter enters the soil below. Red pine thrives in this acid soil, assisted by fungi that extract nutrients from surrounding mineral particles and deliver them to the tree roots through a network of white filaments. But to sustain agricultural crops in this soil, additional organic matter and fertilizer are needed to provide higher levels of available nu-

FELDA SERIES
ARENIC OCHRAQUALF

The color of the faded blue-gray soil shown here, 4 ft. to 5 ft. beneath the surface, is caused by bacterial activity. In this sandy Florida soil, humus stains the surface dark gray. Beneath this, a layer of white sand lies over yellowish-brown soil, colored by limonite and clay.

A sandy soil near The Dells, Wisconsin, has been stained bright red by hematite, an iron oxide mineral. Glaciers ground up iron-ore deposits near Lake Superior and washed them down to this region.

trients. In a cool climate such as that in Minnesota, organic matter decomposes relatively slowly, so once farmers clear the forest and add organic matter to the pale soil, the new organic matter stains the soil dark and that color remains.

A different kind of white soil is found in deserts and other areas where limestone rock is plentiful. Sandy and gravelly soils, formed from the limestone in humid areas, commonly are white and have ample space for roots to grow between the coarse particles. But in some desert soils of southern Arizona and New Mexico, the spaces in the subsoil become filled with calcite, a natural lime cement composed of calcium and carbon dioxide that obstructs root growth and water movement. In climates with abundant rainfall, excess calcite is washed down into the soil out of reach of plant roots, but the sparse rain hasn't leached the lime very far down in much of southern Arizona. When the lime is deposited as fillings in pores of the subsoil, a white hardpan or "caliche" results.

Red and yellow soils—Red and yellow soils are found along the Atlantic Coast from southern New Jersey down to Florida and into Texas. Both colors are produced by iron oxide coatings that cover soil particles or humus. The pigment in red soils is called hematite ("blood" crystal), an iron oxide mineral. The coating on yellow soils is a common mineral called limonite ("yellowish" crystal), composed of both iron oxide and crystalline water (molecules of water trapped in the iron oxide). It takes tens of thousands of years for these bright iron stains to cover the soil particles. Such red and yellow soils usually are found in old landscapes where glaciers and volcanoes have not been actively spreading new soil materials.

Peanuts grow well in many red or yellow soils, such as those in South Carolina, primarily because of the favorable growing season. But organic matter decomposes so fast in this warm, humid climate that even with the constant addition of organic residues, these soils do not turn black.

Bluish-gray soils—Light bluish-gray soils are found in wetlands at a depth of about 2 ft., often beneath deep peat or at the shallow edges of bogs. For centuries, water has saturated these soils almost year round. Bacteria in these subsoils can't get from air the oxygen they need in order to feed on buried organic residue, so they're forced to turn to hematite and limonite in the soil and take oxygen from the iron oxide crystals in these minerals. As a result, the bright red and yellow colors fade to bluish-gray. Most plant roots can't grow here until the excess water has been removed by artificial drainage and air has been let into the soil. These drained wetlands can support yields of crops similar to those obtained from drained muck soils.

If soil were made up entirely of ground-up rock, it would likely be white or off-white, the color of most common rocks when powdered. Instead, an array of natural "paints" (humus and iron oxides) gives soils their rainbow colors and creates visually pleasing landscapes. The changes in this display, sometimes occurring over thousands of years, are part of the fascinating history of the local landscape. We may take great delight in seeing soil colors. They have a beauty all their own. □

Francis D. Hole, an emeritus professor of soil science at the University of Wisconsin-Madison, writes and speaks about soil with the help of songs, puppets and his violin.

The World Under Our Feet

Gardeners can encourage hidden allies in the soil

by Stuart Hill

Illustrations by Michael Rothman

Let's start by getting one thing straight—over 99.9% of all of the life in the soil is beneficial. Without soil creatures, plants would struggle to grow. Large or small, earthworms or microscopic mites, soil animals break down organic matter into forms that nourish plants; they also cultivate and build up the soil, and they control pests and diseases. If there is any shortage of these helpful crea-

tures, we gardeners get stuck with the jobs that they are better equipped to do. We then pay a price in fertilizers, pesticides and toil. I'd rather create conditions that help soil creatures do their jobs, and I think you should, too. I encourage you to become acquainted with our friends in the soil, to explore the next great frontier for gardeners and landscapers—managing the beneficial life in the soil.

You may wonder how I became so passionate about these denizens of the soil. It all started back in the mid-1960's when I was studying the ecology of a cave in Trinidad. The roof was covered with about a quarter of a million bats, so you can guess that the floor was covered with guano. Because most of the life in the cave resided on its floor, I chose to study the creatures that lived there.

Large soil creatures While several foragers recycle organic matter, releasing plant nutrients, another falls prey to a beetle (4X life size).

To avoid being labeled Colonel Bat Guano for the rest of my life, I thought it might be wise to apply my expertise to the soil of gardens, farms and forests.

Life underground

In most parts of the world there is more life beneath the surface of the soil than above it. The underground giants are moles, gophers and other burrowing creatures. Then come relatively large animals such as earthworms and hundreds of species of critters that are not much bigger than the head of a pin, primarily mites (related to spiders), springtails and their relatives. The smallest creatures include nematodes (thin, microscopic worms) and one-celled protozoa. Then there are the fungi, whose threadlike growth spreads through the soil. In a square yard of soil there can be more than ten million nematodes and protozoa, one million mites and springtails, and thousands of other invertebrates.

Let me introduce you to some of my favorite co-workers in the soil. The workhorses are earthworms, especially night crawlers, which are the biggest North American earthworm species. In a healthy soil there can be several hundred earthworms per square yard. If you want to confirm this for yourself, venture out with a flashlight on a night in late spring after a rainy day and you'll see many of them on the soil surface. Earthworms, along with the other larger soil animals, certain insects, slugs and some snails, burrow through the soil, creating channels that improve aeration and drainage. They feed primarily on dead organic matter and on the microorganisms that colonize it.

Another group of my favorites includes mites and springtails. Along with many other small creatures, they wander around in the minute cracks and crevices in the soil (up to their knees in water), browsing on fungi, bits of dead organic matter, an occasional nematode—and one another. There may be hundreds of them in a handful of fertile soil.

The least well-known group, but one that fascinates me, includes the really small guys—nematodes and one-celled protozoa. They swim or slither about in the very narrow water film that covers moist soil particles. Most of them feed on bacteria or on one another. Along with the other groups of beneficial soil organisms, they are essential for the optimal release of nutrients to plants, maintenance of good soil structure and control of plant pests.

The general rule is that the greater the diversity of soil life, the more fertile the soil. If there is a full complement of earthworms in the soil, they can produce as much as 25 tons of worm manure, or castings, per acre per year in a temperate climate. Worm castings are Cadillac fertilizer.

Mites and springtails contribute to soil fertility in a different way. All soil fungi decompose plant wastes, but some fungi also produce antibiotics like penicillin that kill other fungi. In the process of browsing on soil fungi, mites and springtails sometimes temporarily decrease the population of antibiotic-producers, which gives the other fungi a chance to develop and to break down organic matter. And believe me, without these mites, antibiotics can really put a damper on a fungus's dinner party.

I find mites particularly interesting. Picture them wandering along the passages of the soil, browsing on their favorite fungi. The mites eat fungal filaments and spores (equivalent to fungal seeds), but they digest only the filaments. The spores pass unharmed through the mites' guts and are deposited throughout the soil in little packages of waste. Because this waste is fertile, like potting soil, the spores germinate and form new fungal colonies, which then break down organic matter and free up nutrients.

Small soil creatures At 11X life size, soil particles and root hairs dwarf the creatures that live among them.

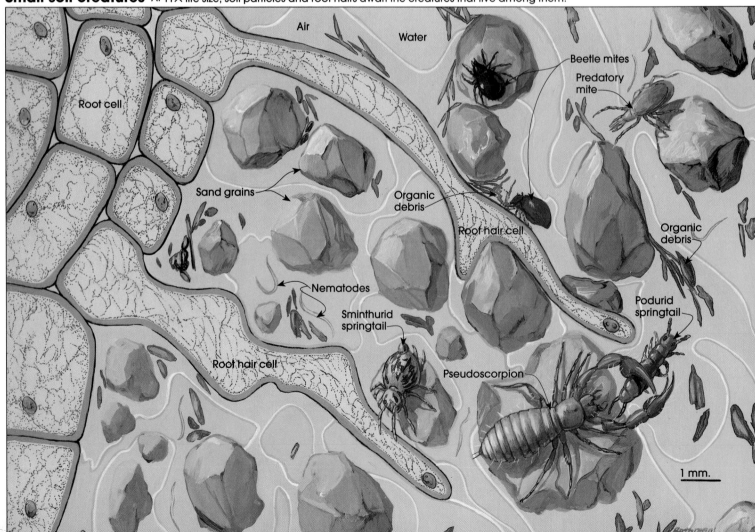

Root cell

Air

Water

Beetle mites

Predatory mite

Sand grains

Organic debris

Root hair cell

Organic debris

Nematodes

Sminthurid springtail

Root hair cell

Pseudoscorpion

Podurid springtail

1 mm.

The mites are essentially practicing a primitive form of gardening. It's humbling to think that this sort of gardening has been going on in the soil for more than 400 million years and that our version is relatively recent. And it's important to realize that the relationships between soil creatures are delicately balanced and can be easily disrupted by human activity.

Gardening in partnership with soil animals

The key to creating a good working relationship with our friends in the soil is to try to see the world from their point of view. Where would you live, how you would get around, what you would eat, and what conditions would make your home ideal for you?

About half of the average soil consists of solid material—mostly mineral particles— and half consists of spaces between the particles. The spaces are half-filled by the film of water on soil particles and half-filled with air. This situation has resulted in the evolution of three primary strategies for living in soil—the earthworm-size animals tunnel through the soil, independent of the air spaces or water film; the mite-size animals live in the air spaces, and the protozoan-sized organisms live in the water film.

Unfortunately, nearly everything that humans do to the soil degrades the living space of soil organisms and kills them, directly or indirectly. The organisms are starved by our habit of not returning all of our wastes to the soil or our tendency to grow the same plants in the same soil year after year, which provides the soil population with a monotonous diet. They are poisoned by pesticides and high-nitrogen fertilizers. When we cultivate and leave the soil bare, soil animals are injured by the rays of the sun and exposed to predators. Desiccation, flooding, fire, compaction from heavy machinery and contamination with a vast range of pollutants are added hazards.

What can we do to provide soil organisms with optimal conditions? First, return organic matter to the soil—wastes such as plant residues, manures and composts. Second, avoid chemical, physical or biological stresses to the soil. Don't poison the soil with toxic pesticides. Cultivate as little as possible. To maintain the layered structure of the soil, loosen it, rather than turning it upside down. Most soil animals live in the top three inches, and if you turn the soil upside down, it is like removing the penthouse suite and putting it underground—the occupants will not be pleased.

To help the soil life in my garden, I spread compost on the surface of the soil in the fall after I clean up the garden, loosely fork it in, and lay down a thin mulch of leaves or hay to protect the soil surface during the winter. In the spring, I rake back any remaining mulch and, with minimal disturbance, prepare the soil. To provide a balanced range of nutrients, I add compost and any other needed amendments to the planting holes. (I use only naturally occurring materials, such as bonemeal and kelp.) I sprinkle sieved compost on top of my seed beds to nourish and protect them. Once the soil warms up and the plants are established, I rake the mulch back onto the bed. During the growing season, I keep the soil moist, but I don't flood it.

There are many functions performed by the creatures in the soil—many more than we are fully aware of, and many more than I've been able to touch on in this article. Remember that whenever we eliminate one of these organisms, we inherit its job, one at which it is an expert and we are, at best, novices. By conserving our allies in the soil, we help them support us and the rest of the planet. We also make it possible for more young idealists to one day fall in love with the amazing inhabitants of the soil and enjoy their goings-on as much as I have. □

Stuart Hill, a professor of entomology at McGill University, Montreal, Canada, is director of Ecological Agriculture Projects, a resource center for sustainable agriculture.
Michael Rothman is a natural-science illustrator who lives in Ridgefield, Connecticut.

Very small soil creatures Air spaces and films of water around soil particles weave a home for tiny soil creatures (44X life size).

Home Soil Testing

Are kits worth the effort?

by Mark Kane

I started gardening on a worn-out farm with acidic clay soil. In two years, with truckloads of ground limestone, barn bedding and manure, and heavy mulches of straw and hay, I had workable soil and healthy plants. I often wished I had started with a soil test, and then retested at intervals to see what my labors had wrought. I was curious about soil-testing kits, but had never bought one. Recently, after a move and a new gardening start on acidic sandy soil, I had a chance to satisfy my curiosity.

I looked at products that test nitrogen (N), phosphorus (P) and potassium (K), the nutrients plants need in greatest quantities, and also test pH, a measure of soil acidity or alkalinity. I bought the three kits I could find that seemed, by price and availability, to be aimed at gardeners: LaMotte's Model EL, Luster Leaf's Rapitest Soil Test Kit, and Sudbury's Model D. The latter two are sold at many garden centers and by some mail-order suppliers. All three manufacturers sell their kits by mail (see Sources, p. 20).

The kits are straightforward to use.

They include similar basic components—test chemicals, graduated containers, instructions—and demand similar procedures. In general, you combine measured amounts of soil and a liquid test chemical, shake them together, filter the solution or let it settle, then compare the color of the solution with a chart to read the result. For some tests, when the solution settles you add one or more chemicals to produce the indicator color. If you have any form of color-blindness, you'll need help reading the results—the color charts cover the spectrum.

I spent three days working with the kits, evaluating their ease of use, comparing their instructions and manuals, analyzing soil samples, compiling the results, and calculating fertilizer recommendations. I also sent portions of the soil samples for testing to the Connecticut state lab. (The results appear in Chart 1 on the facing page.) After all this, I found myself unable to form tidy conclusions. It seems likely to me that your temperament and approach to gardening will have a large say in whether you're attracted to kits, state-lab tests or testing at all. Soil testing is a complicated subject, as the following experiences point out.

Sampling and testing

The first step in soil testing is collecting representative soil samples. Since the kits

test minute portions of soil, roughly the volume of a pea, collecting a representative sample is important. All three kits give generally adequate instructions for making a sample—take slices of soil from 2 in. to 6 in. below the surface at several closely spaced locations and mix them together thoroughly.

I collected five soil samples from three different sites; Chart 1 shows three samples to save space. Numbers 1 and 2 came from soils amended for vegetable gardening in the preceding six months with limestone, fertilizers and organic matter. Number 3 came from a plot of freshly tilled turf. The kit manufacturers advise sifting or screening the soil, so I passed each sample through a square of window screen. This excluded a lot of organic matter—by volume, between 10% and 30% of the samples. Organic matter supplies nitrogen to plants, but none of the kits suggest accounting for it.

Working with the kits for three days gave me a good deal of respect and sympathy for lab technicians and people who wear bifocals. In one fashion or another, each kit demands finicky, tedious attention. Of the three kits, I found LaMotte's the easiest to use. The test tubes are large and the solutions settle quickly. The indicator solutions are relatively easy to read, largely thanks to their clarity—you trans-

CHART 1: COMPARING THE SOIL TESTERS

		Sample 1				Sample 2				Sample 3			
		pH	N	P	K	pH	N	P	K	pH	N	P	K
Connecticut State Laboratory	analysis	5.5		VL	M	4.8		M	VH	5.4		MH	VH
	recommendation	7 lb.	3.2 oz.	6.4 oz.	6.4 oz.	11 lb.	3.2 oz.	3.2 oz.	3.2 oz.	7 lb.	3.2 oz.	3.2 oz.	3.2 oz.
LaMotte Model EL	analysis	6.0	L	L	MH	4.5	MH	L	VH	5.0	MH	ML	VH
	recommendation	2.5 lb.	6.4 oz.	8 oz.	5.6 oz.	9.9 lb.	4 oz.	8 oz.	2.4 oz.	7.4 lb.	4 oz.	7.2 oz.	2.4 oz.
Luster Leaf Rapitest	analysis	6.0	L	L	H	4.5	L	L	L	4.5	L	L	L
	recommendation	3.9 lb.	3.4 oz.	5.3 oz.	1.3 oz.	16.9 lb.	3.4 oz.	5.3 oz.	5.3 oz.	16.9 lb.	3.4 oz.	5.3 oz.	5.3 oz.
Sudbury Model D	analysis	6.0	5	2.5	5	5.0	25	5	3	6.0	25	2.5	3
	recommendation	2.3 lb.	3.2 oz.	3.2 oz.	6.4 oz.	6.8 lb.	1.6 oz.	1.6 oz.	7.2 oz.	2.3 lb.	1.6 oz.	3.2 oz.	7.2 oz.

NOTES: All but Sudbury express analyses in broad ranges (VL for very low, L for low and so on). Sudbury's figures are in parts per million of soil.

The kits measure pH in 0.5 increments; the state lab measures pH in 0.1 increments. The manufacturers' recommendations indicate the amount of ground limestone needed to produce pH 6.5 on 100 sq. ft. of soil. The lab's recommendation is for pH 6.4.

The N, P and K recommendations are meant to produce good soil fertility. They are given in actual quantities of the elements to be applied per 100 sq. ft. of soil. For a given fertilizer, you must know the percentage of each element, and adjust the amount accordingly. The state lab feels that there's no reliable test for nitrogen. Their N recommendation is based on experience and the P and K results, and represents the least amount of nitrogen that will produce good growth for the specified crop.

The author used three moderately priced soil-testing kits (facing page). From left to right are the LaMotte Model EL, Luster Leaf Rapitest, and Sudbury Lawn & Garden Model D. In front of the kits are some of their components: color charts, filter disks, tin wires and the sandpaper used to brighten the wires for phosphorous tests.

fer reagents from settled test tubes to clean test tubes, using large eyedroppers, before adding other reagents to produce indicator colors. LaMotte supplies seven test tubes and three eyedroppers, which allows you to test a soil sample for N, P, K and pH without stopping to clean equipment.

Luster Leaf has a unique method for filtering solutions of soil and reagent. The kit includes a small syringe body and filter paper die-cut into disks. You put a disk inside the syringe body, covering the outlet, then add a small amount of soil and a measured amount of reagent. After shaking the solution and letting it settle, you use the syringe plunger to force the solution through the filter disk into a test tube. Don't rush the settling. I did, and clogged the filter paper with small soil particles, which forced me to start again. Luster Leaf supplies only one test tube and syringe body, so you have to wash and dry them three times to test a soil sample for N, P, K and pH.

Sudbury's kit uses no filtration. You put a bit of soil in a small test tube, add reagent up to a graduation on the test tube, shake, let the solution settle, and read the indicator color. The P, K and pH tests settled within a few minutes, but the N test took hours. To do N, P, K and pH tests for one soil sample requires five test tubes. Sudbury supplies four, obliging you to

Following the kits' instructions, Kane screened soil samples before testing. For some samples, screening removed considerable quantities of organic matter, which contributes nitrogen and other elements to the soil. The kits don't suggest adjusting test results to account for organic matter. Here the unscreened portion (on the left) has many visible leaf fragments; the screened portion has none.

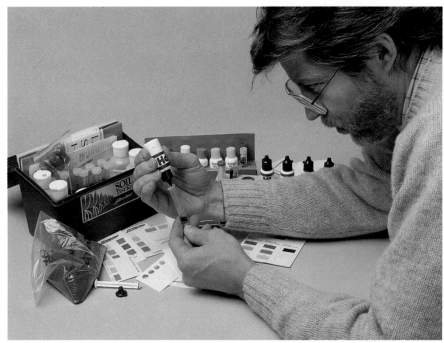

The three soil-testing kits the author used demand similar procedures. You mix a small amount of soil and reagent to produce a colored indicator solution. Measurements are exacting. Here Kane adds reagent to 0.5 ml of soil, filling Luster Leaf's container, a graduated syringe, to the 2.0 ml mark.

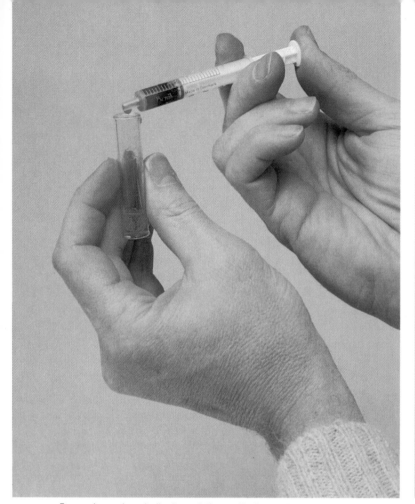

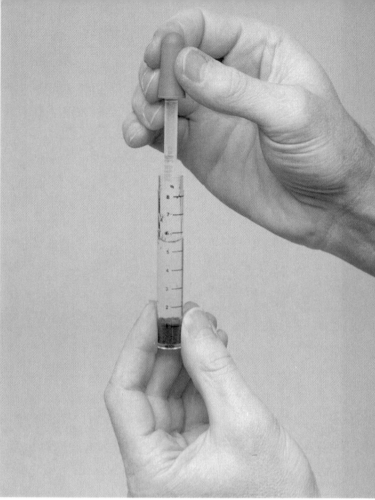

To produce clear indicator solutions with the Luster Leaf kit, you filter the mix of soil and reagent (above left). A disk of filter paper at the nozzle of the syringe body traps the soil. In LaMotte's large test tubes (above right), the mix of reagent and soil settles quickly. Once it has settled, you draw off the clear solution with a pipette, put it in a clean test tube and add other reagents to produce the indicator solution.

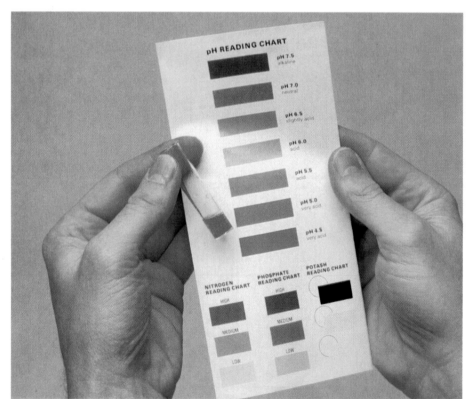

To read the test results, you compare the indicator solution to a color chart. Here Kane holds a Luster Leaf pH test beside the pH chart. The reading is pH 4.5, indicating that the soil sample is highly acidic.

wash one to finish the cycle. After a few washings, the printed graduations on Sudbury's test tubes grew faint. Before they could vanish, I went over them with a knifeblade, scoring the plastic. After five washings, one test tube had no sign of printed graduations. LaMotte's test tubes and Luster Leaf's syringe showed no wear.

I could dispense reagents from the LaMotte and Luster Leaf containers with precision, thanks to built-in nozzles, but not from Sudbury's plastic bottles, which have a lip inside the mouth. No matter how gradually I tilted the bottles, the reagent bulged on the lip and then let go like an avalanche. Several times, when I filled a test tube to within a few drops of the proper graduation, I let well enough alone, at some cost in accuracy. After testing five soil samples, I had used up the reagent for the P test, and had very little left for the other tests. The kit is described by Sudbury as providing enough reagent to test eight samples.

I had trouble comparing some indicator solutions to the color charts. The Sudbury color chart gives three shades of orange for pH 4.5, 5.0 and 5.5. With one sample, I asked several people to do the comparison. Four of them said pH 4.5, one said pH 5.5 and I said pH 5.0. The Sudbury N test solution also gave me problems. After hours of settling, it was still slightly turbid and so dark that the color was hard to see. The instructions tell you to hold the test tube in good light; I had to hold it up to a window. I called Sudbury, and they suggested that I strain the mix of soil and reagent through a paper napkin or a coffee filter to get a clearer solution.

I found LaMotte's P test difficult to interpret. You add two drops of indicator to the solution, swirl the mix, and look for a color change from purple to blue. If there's no change, you repeat the process. The number of drops is a measure of the phosphorous in the sample, with ten drops or less indicating a very high phosphorus level, and a greater number of drops indicating a lower level. I had trouble deciding when the purple changed to blue. On occasion, the top ¼ in. of the mix turned clear or pale blue and then turned blue or a deeper blue after I added two, four or six more drops of indicator liquid.

I wondered how my procedures affected the results. Since the manufacturers didn't outline standard procedures, I invented my own: I didn't tamp or jiggle the soil when measuring, and I added reagents until the fluid just touched the appropriate graduation. It seemed to me that big variations were possible. For example, when I added soil samples to Luster Leaf's syringe and Sudbury's test tube, the dirt tended to clump and leave air pockets. When I added reagent, it sometimes perched on the dirt and failed to fill the air pockets. After mixing had released the trapped air, the solution of reagent and soil sometimes decreased 10% in volume. As an experiment, I did a P test with each kit, adding 10% more soil than called for, and 10% less reagent. This altered the results of the Sudbury and Luster Leaf tests, and increased the number of drops that produced blue in the LaMotte test.

I asked all the manufacturers about the safety of their kits. They characterized the reagents as relatively weak chemicals, and said that the tests could be safely flushed down a sink drain. All recommended that as a matter of course the reagents be kept out of the reach of children. Sudbury's pamphlet makes no mention of these points. Luster Leaf's literature discusses safety but not disposal (and has the good sense to suggest that "it is educative for responsible children to carry out the tests"). LaMotte's information package thoroughly covers safe use and disposal. It recommends wearing safety glasses during testing; notes that its reagents are listed on Poisindex, a computer database that serves most local poison-control centers; and includes Material Safety Data Sheets, which identify the reagents and describe proper disposal methods.

Test results

Chart 1 shows the results of my testing, and the recommendations made by the

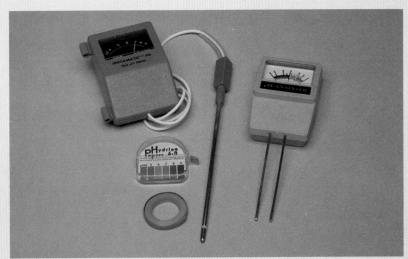

Kane used these three pH testers. At bottom left is Hydrion's roll of litmus paper with plastic dispenser and color chart. Above it is the Instamatic meter, with its 6-in. probe and 25-in. lead. On the right is the pH Analyzer, a meter with two 3½-in. metal probes.

pH testers

When I started a new garden recently, I felt sure that turning under 6 in. of shredded leaves in the fall and adding a bit of 10-10-10 in the spring would provide my tomatoes and beans with an adequate supply of N, P and K, but I wondered what to do about pH. It's common knowledge that soils here are acidic, so I spread roughly ½ lb. per square foot of ground limestone with the shredded leaves and hoped. It was not enough, as I learned from a soil test the next summer. The pH was 5.5.

The experience prompted me to look for products that measure soil pH alone. I found three: the pH Analyzer, the Instamatic Soil pH Meter, and Hydrion papers (see Sources, p.20). The first two are gauges; the third is a roll of litmus paper.

There's a direct relationship between pH and the electrical current that naturally flows through damp soil between two metals. The gauges detect this current and read out the pH. You insert the pH Analyzer's two thin 3½-in.-long probes into damp soil, and read the pH on the gauge. The Instamatic Soil pH Meter has a single thin probe hooked up to a gauge by a 25-in. length of insulated wire. The probe has three parts: a shank of one metal, a tip of another metal, and an insulating spacer between them. You insert the probe about 5 in. into the soil, and the gauge indicates the pH reading.

Hydrion litmus paper is equally straightforward to use. You press a piece of the paper against a damp soil sample for 30 seconds, then match the color of the paper with one of six colors that indicate pH readings from 4.0 to 9.0. The roll is 180 in. long, enough for 90 pH tests, at a cost of pennies each. The manufacturer told me that the paper will remain accurate for five to seven years, provided it's protected from dampness.

Chart 2 (below) shows the results I obtained with the pH testers on the same three soil samples shown in Chart 1. I repeated the tests three times, and got the same results each time. Then I followed the manufacturers' instructions for sandy soil and calculated the quantity of limestone needed to raise the pH of 100 sq. ft. of soil to 6.5. The differing pH readings may have to do with the sensitivity of the gauges to salts dissolved in the soil. Most of the variation in recommendations can be accounted for by the manufacturers' disagreeing about how ground limestone alters pH. Hydrion says that 40 lb. spread on 1,000 sq. ft. of sandy soil raises the pH one point, while Instamatic says 52 lb. The pH Analyzer, inexplicably, says that 3.5 lb. raises the pH from 6.0 to 7.0 on 1,000 sq. ft. of sandy soil; this quantity is roughly ten times too low. —M.K.

CHART 2: pH TESTERS

		Sample 1	Sample 2	Sample 3
Hydrion papers	analysis	5.5	5.5	5.0
	recommendation	4 lb.	4 lb.	6 lb.
Instamatic Soil pH Meter	analysis	6.5	4.5	5.6
	recommendation	0	10.4 lb.	4.7 lb.
pH Analyzer	analysis	6.5	6.2	6.0
	recommendation	0	0.1 lb.	0.2 lb.

three manufacturers and the Connecticut lab. All but Sudbury express the amount of nutrients found in the soil in plain English: low, high and so on (LaMotte and the state lab in five increments, Luster Leaf in three). Sudbury gives nutrients in parts per million of soil, which may sound more precise, but they don't state whether the numbers represent total nutrients, available nutrients or some other measure. I followed each manufacturer's instructions to calculate their recommendations, the amounts of N, P and K that would produce good soil fertility. For comparison, I translated all these into N, P and K quantities per 100 sq. ft.

As Chart 1 shows, the results and recommendations vary widely, without consistent patterns. You might conclude that one set of results must be right, or nearly so, and the others wrong. Unfortunately, it's not that simple. Remember that soil tests measure a portion of the N, P and K in a soil sample, and the sample's pH. They do not say how much N, P and K plants need, or how to alter pH. They are yardsticks, not prescriptions. The crucial point is that soil scientists use many yardsticks—different ways of measuring the amount of a nutrient or the soil pH, each giving a different result. I think the manufacturers use different yardsticks, hence the variation in analyses. As yardsticks should, the kits gave reproducible results—each agreed with itself when I did repeated tests of the same soil sample. Results alone are meaningless, however. What counts for gardeners are recommendations.

Converting test results into recommendations for adding fertilizer and altering pH is complicated. The number of factors to consider is daunting. Plants on sandy soil respond differently to fertilizers and pH than do plants on clay soil. The amount and condition of organic matter in the soil influence test results and plant growth. Climate plays a role. Thanks mainly to a long warm season, a southern garden may produce good-quality vegetables in soil that has half the phosphorous a northern garden would need. And different forms of fertilizers differ in their effects. Some concentrated nitrogen fertilizers, such as ammonium nitrate and urea, which are common components of commercial mixes, release a large percentage of their nitrogen in one season. Being strongly acidic, they also lower pH. Other nitrogen fertilizers, such as manure and compost, release their small store of nitrogen over several seasons, and may raise pH slightly. The various forms of phosphorus, potassium and lime show similar differences.

Making recommendations demands a solid grounding in experience. Only one group I know of, the land-grant universities, has done the necessary work. For decades, researchers at these universities have established test plots on the major

soils of each state. They've grown different crops with varying amounts of N-, P-, K- and pH-altering amendments; recorded the health and yield of the crops; and made soil tests. In this welter of variables, they've found patterns that allow them to translate test results into soil-amendment recommendations that work for the crops, climate and soils of their area.

The kit manufacturers simplify these translations to produce an all-purpose recommendation and give users a clear course of action. It would hardly be possible for a kit sold nationwide to give guidelines for each state, let alone every soil and climate. I think different approaches to simplification account in large measure for the kits' different fertilizer recommendations.

The incongruity in pH recommendations also stems from simplification. The kits measure pH in 0.5 increments, obliging you to round off the test readings. The result is room for error. The manufacturers also differ about amendments that alter pH. Sudbury says that 45 lb. of limestone per 1,000 sq. ft. of sandy soil will raise pH one point; LaMotte says "slightly less than" 55 lb. (which I took to mean 10% less, or 49½ lb.); and Luster Leaf says that the amount varies, depending on the initial pH (by my calculations the figure averages 29½ lb.). It's possible that the manufacturers have different forms of limestone in mind, but the kits say nothing about the subject. The manufacturers also disagree about lowering pH. To lower pH from 7.0 to 6.0, LaMotte recommends applying

SOURCES
Test kits
Note: Kit prices reflect the number of soil samples each kit will test.

LaMotte Soil Test Kit Model EL: LaMotte Chemical Products Company, P.O. Box 329, Chestertown, MD 21620. 301-778-3100; 800-344-3100 for direct orders. $28.90 plus shipping.
"LaMotte Soil Handbook," Code 1504, $1.55 postpaid (ATTN: Margaret Phillips).

Luster Leaf Rapitest Soil Test Kit: Luster Leaf Products, Inc., P.O. Box 1067, Crystal Lake, IL 60014. 815-455-5160. $16.09 postpaid.

Sudbury Soil Test Kit Model D: Farnam Companies, Inc., P.O. Box 34820, Phoenix, AZ 85067. 800-528-1378. $13.99 postpaid.

pH testers
Hydrion papers: Micro Essential Laboratory, Inc., 4224 Avenue H, Brooklyn, NY 11210. 718-338-3618. Soil Test Kit #55-S, $5.00 postpaid for 180-in. roll.

Instamatic Soil pH Meter: AMI Medical Electronics, Inc., P.O. Box 148, Ronkonkoma, NY 11779. 516-348-0300. $20.00 postpaid.

pH Analyzer: Environmental Concepts, 710 N.W. 57th St., Fort Lauderdale, FL 33309. 305-491-4490. $23.45 postpaid.

alum (aluminum sulfate) at the rate of 1 lb. per square yard, Sudbury advises a little more than ¹⁄₁₀ lb. per square yard, and Luster Leaf suggests quantities without saying how much they will lower pH.

While I think I understand why the manufacturers and the state lab give different recommendations, I don't know how to choose among them. Fortunately, it's possible to err widely on N, P, K and pH and still grow satisfactory plants, especially on clay soils and those with a fair amount of organic matter. Sandy soils are less forgiving. Though the recommendations on Chart 1 differ considerably, I suspect that none of them would harm my garden, because I've added so much organic matter. At worst, they might waste a few dollars of fertilizer. I conclude, however, that gardeners ought to regard soil-test recommendations as points of departure, not guarantees of soil fertility.

Summing up
I think it's possible to use a soil-testing kit profitably, provided you emulate the land-grant universities and conduct your own field trials. Make regular tests, follow the manufacturer's recommendations, observe the response of your plants closely and keep detailed records. If the plants show nutrient deficiencies, alter the recommendations until the plants do well. Eventually you should know how to read the test results and adjust the recommendations to suit your conditions. LaMotte, to its credit, suggests that this course is the proper way to use its kit.

If I wanted to undertake regular soil testing, however, I would probably use my state lab. Compared with fees in some states, Connecticut tests, at $2 each, are a bargain. I dropped by the local Extension office, paid $10 for five sets of mailing envelopes, filled out the enclosed questionnaires, mailed off the samples and was done. To test five samples with the fastest kit would have taken me at least 2½ hours, and nearly the same amount of money. (Per-sample costs range from $1.45 to $1.75 for the kits; less when replacement reagents are purchased.) The state lab makes recommendations that take into account the crop, organic matter and soil texture, and that are based on field trials. If I have a question, I can phone the lab.

If you have a well-developed eye, though, I think you can largely dispense with soil tests. The yellowing leaves of your corn plants will tell you they need nitrogen; the yellowing leaves of your rhododendrons will tell you they lack iron, and need a lower pH. Keeping good records of what you add to the soil and how your plants respond is likely to teach you as much as soil testing will. □

Mark Kane is an associate editor at FG.

Evaluating Potting Soil

What makes a good mix?

by Rita Buchanan

Whether you're growing African violets on the windowsill, cherry tomatoes on the patio or fuchsias in a hanging basket, to grow plants in pots you need potting soil. Chances are you can't just dig it from the garden—it's a rare achievement to have garden soil that's crumbly and porous enough to provide good drainage in a pot. More than likely you'll need to make or buy some.

Old-time gardeners made their own potting soil. The recipe was simple: Mix together one part loamy garden soil, one part compost or peat moss, and one part coarse sand. Homemade potting soil is good stuff, like home-baked bread, but busy modern gardeners often rely on store-bought products. Any local garden center has an assortment of bagged products labeled "potting soil." The question is, how do you know which to buy? Unlike food products, potting soils don't carry lists of ingredients on the label. You probably won't know what you've bought until you open the bag at home.

Some bags include nothing but soil—often a dark-colored muck that would be hard to work with in a garden, let alone in a pot. This kind of soil is usually bagged and distributed on a local scale. By contrast, most professional-grade, nationally marketed modern potting soils don't contain any *real* soil at all. They're "soilless" mixes, generally composed of peat moss, ground pine bark, and vermiculite and/or perlite, with lime and nutrients added (see the sidebar on p. 24 for more about these components). Soilless mixes were developed in the 1950s and '60s when commercial nurseries were beginning to grow more crops in containers, and couldn't reliably get good loam to use in soil mixes. As the idea of using soilless mixes caught on, an industry developed to supply ready-mixed products both to nurseries and to home gardeners. Soilless mixes have their pros and cons, which I'll discuss later; for now, the point is

To measure the porosity of potting soil, add water, then drain it off. With an ideal sample, you can add about two cups of water to a quart of dry soil (above). A few minutes later, you can drain off about one cup of water (below). The volume of water that drains off equals the amount of air space in the soil.

that they dominate the ready-mix market.

The proliferation of mixes can be confusing. Some gardeners prefer to use special mixes for different kinds of plants. I make custom mixes when I'm feeling solicitous toward favorite plants, but I really don't think it's necessary. Commercial growers know you can pot a majority of plants in the same kind of soil and grow them successfully. This ideal mix is defined not by its list of ingredients—several formulas approach the ideal—but by its physical and chemical characteristics.

Whether you buy bagged potting soil or prepare your own, there are advantages to using the same mix for as many of your plants as possible (I make exceptions for orchids, bromeliads, cacti and tiny seedlings). Standardizing the soil simplifies your watering, fertilizing and repotting routines. And buying larger quantities of one mix is more economical than choosing several small bags of different mixes.

The ideal potting soil— a measure for the mixes

Whenever I try a new mix, commercial or homemade, I judge it against the set of standards that defines my ideal. This is a shortcut compared to potting up a few plants and watching their response. Assessing the physical characteristics of a soil mix is fairly straightforward—I do a series of simple tests that use common equipment and common sense. It's harder to evaluate a soil mix's chemical characteristics at home—that's what soil-test labs do—but there are some steps you can take. Here's what I look for.

Porosity—One of the first things I check is porosity. In a recently watered pot of ideal potting soil, only one-half the volume will be filled with particles of mineral and organic matter. One-quarter of the volume will be filled with water, clinging in the small pores between particles, and one-quarter of the volume will be filled with air, which enters the larger pores as water drains out.

I determine porosity by filling a jar with dry soil, measuring how much water can be added before it overflows, and then tipping the container and measuring the volume of water that drains off within a few minutes. The volume of water that drains off equals the volume of air in the soil; the difference between the amount added and the amount drained off is the net volume of water in the soil. Ideally, you can add about two cups of water to a quart (four cups) of soil, then drain off one cup. In practice, some mixes accommodate more than two cups of water, because some is absorbed into the particles, as well as clinging to the spaces between particles. As long as the minimum one-cup's worth of air space is available, it does no harm if the mix retains extra water.

Particle size—The next thing I check is particle size. Ideal potting soil doesn't have any large clods, pebbles, sticks or chunks of debris, and it doesn't have any dust-fine particles either. I usually screen a sample with a ¼-in.-mesh sifter. Chunks that don't fit through the mesh are too big. Then I put the sample in a sifter with ⅟₁₆-in. mesh. Tiny bits that pass through the fine mesh are too small. I'm satisfied if most of the particles fit the ¼-in. to ⅟₁₆-in. size range. A few larger chunks don't do any harm if the potting soil will be used to fill 6-in. or larger pots. But I don't like to see more than a dusting of particles pass through the fine-mesh screen. Particles that are too small sift into the spaces between larger particles, creating a densely packed, nonporous mix. Fine, silty sand is a frequent culprit in this regard. Over time, ingredients may decompose into smaller pieces that fill pore spaces and increase a mix's density; this breakdown is one reason not to recycle used potting soil.

Response to watering—It's easy to water a pot of ideal soil. The water doesn't bead up on the surface. It penetrates quickly, and much is absorbed as it flows through the pot. The excess drains out promptly. If water moves very slowly, it means that the mix's pores and particle sizes are too small. Don't try to judge peat-based mixes as they come from the bag, though; they're very dry and hard to wet initially. Add hot water and wait a few hours for it to penetrate the peat. Subsequent routine watering isn't difficult unless you let pots filled with these mixes get too dry. Some commercial mixes include chemical wetting agents that make it easier to wet the peat moss, but these chemicals are reported to damage the root tips of sensitive plants.

To measure compactibility, observe what happens when you add water to or press down on a sample of potting soil. Both these samples started out the same volume. Ideally, a sample should resist compression (on the left). Be careful with fluffy mixes that collapse under pressure (on the right)—once compacted, they hold too much water and too little air.

Most of the ingredients that are used in soilless mixes can absorb their own weight in water—they weigh twice as much, or more, when wet as when dry. Mixes containing soil or sand may take up as much water, but they're much heavier to start with, so the difference isn't as marked. I like to weigh a potful when it's dry and again when it's wet, but you can simply lift a pot to feel the difference. If you stick with the same mix for all your plants, you'll soon be able to judge a pot's water needs just by hefting it.

An ideal potting soil remains evenly mixed and homogeneous over time, but there's a tendency for repeated waterings to stratify the contents of a pot. Some ingredients—Styrofoam, perlite and sometimes flakes of bark—are so light that they float to the surface. This tendency is exaggerated if the mix includes smaller particles of denser materials such as fine sand, soil or peat. The small heavy particles settle into a denser layer at the bottom of the pot that tends to waterlog, while the large light particles form an arid zone at the top. I water a sample pot of mix and check to see what rises and what, if anything, flows out the bottom. The less a mix stratifies, the better.

Volume stability—Ideal potting soil doesn't shrink or swell; it fills the same volume wet or dry. I wet a sample, and check to see if it cracks or pulls away from the pot as it dries. It's hard to water soil that shrinks into a dry block, leaving a gap around the edge of the pot—water just flows through the crack without soaking into the soil. (To restore a shrunken soil mass to full volume, soak the entire pot in a bucket of water for an hour or so.)

Any soil mix will fluff up a little when you mix it and will settle down when you use it to pot a plant, but this change in volume should be minimal. It's hard to position a plant at the right depth in a fluffy mix that loses up to a third of its volume when compressed. Also, compressing a mix closes the pores that should ideally be open to air. I gently squeeze a damp sample to test it for compactibility. If it can be compressed by squeezing, it will eventually be compressed by watering and gravity. Handled gently, fluffy mixes are acceptable for short-term use, such as for germinating seedlings. For plants that will be in a pot several months or longer, I choose a mix that resists compression.

Freedom from contaminants—Potting soil should be free from insects and other arthropods, weed seeds, and disease-causing organisms. The peat moss, bark, perlite and vermiculite used in soilless mixes are more or less sterile when they're bagged, but any mix is subject to contamination once the bag has been opened.

Sand and topsoil are the ingredients

To measure particle size, sift a sample of soil through ¼-in.- and ⅟₁₆-in.-mesh screens. Ideally, most of the particles will pass through the coarser mesh (on the left), but not through the finer mesh (on the right).

Ready-mixed potting soil offers convenience, but how do you decide which kind to use? Author Buchanan outlines a way to compare any mix with a set of ideal standards.

Bark

Sphagnum moss peat

Reed-sedge peat

Vermiculite

Perlite

Sand

What's in soilless mixes?

Most of the commercial ready-mixed potting soils available today have been developed from formulas originated at Cornell University and the University of California, and combine two or more of the ingredients described below. The label on a bag of potting soil rarely specifies the identity and proportion of its ingredients, but once you open the bag, you can easily recognize the major components.

The ingredients in soilless mixes may function to retain nutrients, but none are a major source of plant nutrients. Some soilless mixes are amended with an initial supply of fertilizers and trace elements; most include some form of lime, to balance the pH.

Sphagnum moss peat—An abundance of terms—sphagnum moss, sphagnum peat, peat moss, peat, Canadian peat moss, and reed-sedge peat—refer to a group of products mined from boggy areas. Sphagnum moss is a rather coarse-textured moss that forms wide, flat colonies in freshwater bogs. The plants are slow-growing perennials that form a half-inch or so layer of fresh green growth each year. The old growth, lower on the stems, gradually darkens from yellowish-tan to brown, and eventually to a dark brownish black, as it's compressed and decomposed over time. Old, dark, dead sphagnum moss is called peat moss, sphagnum peat, or simply peat. Canadian peat moss is simply peat moss collected in Canada, as opposed to Scandinavian or continental U.S. peat moss. Reed-sedge peat is different stuff—it's dead reeds, sedges, cattails and similar marsh plants. If relatively young, it's coarse-textured with lots of visible stems. It quickly decomposes into a fine-textured, dense, humusy muck. Although inexpensive, reed-sedge peat is less desirable than sphagnum moss peat.

All forms of sphagnum moss and peat are acidic; limestone is added to counteract this acidity in a potting mix. All forms of peat absorb and retain water well. Younger, less-decomposed, lighter-colored, coarser-textured products provide better aeration than do older, darker, fine-textured peats, which have few large pores for air to penetrate.

Bark products—Using tree-bark products in soilless mixes solves two problems: it's a less expensive alternative to peat moss, and it makes use of an otherwise wasted by-product of the timber industry. Bark-based mixes are especially popular in southern, Rocky Mountain and Pacific states. Pine bark (from several species of pines) is most widely used, but some use has been made of hardwood bark and redwood sawdust.

Before it's added to a potting-soil mix, any kind of bark product is ground into ½-in. or smaller chips. Most processors then compost the bark, supplementing it with nitrogen to stimulate microbial activity. Recent research suggests that composted pine bark has anti-fungal properties and that plants potted in bark-based mixes are less susceptible to root rots. Bark products do not absorb or retain water or nutrients as well as sphagnum moss or peat products do. Bark-based mixes tend to provide better aeration than peat-based mixes do, but dry out more quickly. Lime is added to counteract the bark's acidity.

Other composts—Other kinds of composted plant products are finding their way into potting-soil mixes in some parts of the country. These include feedlot manure, peanut hulls, sugar-cane stems and other agricultural by-products; and leaves, yard wastes or sewage sludge composted by municipalities. Most of these products are newly available and haven't been widely tested, but they show promise.

Perlite—Perlite is a glassy, white volcanic rock that's been crushed and heated to about 1800°F. Heating makes the rock expand like popcorn; as a result, perlite is very lightweight. It comes in small, irregular, sharp-edged particles that don't compact or decay. Water clings to the surface of the particles, but isn't absorbed by them. The pH is around neutral, and nutrient value is insignificant.

Vermiculite—Vermiculite is produced by heating mica, a mineral that's naturally layered like the pages of a book, to a temperature of 1400°F. Heating causes the layers to separate, and the result is a lightweight puffy material. Handled gently, vermiculite provides plenty of air spaces in a mix, but, unfortunately, if you press down on wet vermiculite, you can easily squeeze it into an irreversibly dense, compact, waterlogged mess.

A measure of dry vermiculite weighs about the same as dry perlite, but the vermiculite can absorb more water and nutrients. Vermiculite supplies potassium and magnesium, essential plant nutrients, and has a near-neutral pH.

Styrofoam—White beads, usually ⅛ in. to ¼ in. in diameter, of Styrofoam plastic are used in some commercial potting mixes as a substitute for perlite. Styrofoam serves as a space filler, and little more. I can't think of any advantages to using Styrofoam; a major disadvantage is its tendency to float to the surface of a pot, rather than stay integrated in the soil mix.

Sand—There's sand, and there's sand. Horticulturists debate the value of rounded vs. angular particles, quarry sand vs. dune sand, and so on. I don't think any of that matters so much as particle size. The coarser, the better. Fine sand settles into the spaces between other ingredients and makes a dense mix that excludes air. When I make my own mixes, I only use sand that's too coarse to pass through a window-screen sieve (¹⁄₁₆-in.-mesh). Clean, washed sand has a near-neutral pH and little if any nutrient value. Sand is much heavier than any other ingredient used in potting soils. The added weight is an advantage for tall, top-heavy plants that might otherwise blow or tip over, but it's a disadvantage when you have to carry or move the potted plants.   —R.B.

most likely to carry contaminants to a soil mix. Heating damp soil to a temperature of 180°F for 30 minutes will kill most pathogens. You can buy small electric steam pasteurizers from greenhouse-supply dealers, or heat soil in your oven or microwave. To be honest, I never pasteurize the sand, topsoil or mixed potting soil that I use at home, and so far I've had only a few minor problems—occasionally seedlings damp off, annual weeds sprout up, or earthworms hatch inside a pot.

Soil pH—The recommended pH range for potting soils that contain real soil is about 6.0 to 6.5, and for soilless mixes about 5.5 to 6.0. I wet a sample with distilled water and use litmus paper to get an approximate pH reading.

The pH of a soil mix in a pot is likely to change over a period of months in response to the application of soluble fertilizers, and to watering with acidic or alkaline water. (Water isn't necessarily neutral. My well water here in Connecticut is an acid 4.5; in Texas where I used to live, the tap water was an alkaline 10.0.) Many plants can tolerate any soil pH between 5.0 and 8.5 if supplied with fertilizer that includes all the major, minor and trace elements, but beyond that range plants often show signs of nutrient deficiency. Adding small amounts—about 1 tsp. per gallon of soil—of dolomitic limestone or soil sulfur will raise or lower the pH, respectively. Sprinkle the material on the surface of the soil in a pot and water in well. It may take a week or more before the effect is noticeable.

The ideal potting soil maintains its original pH over time; it has a high buffer capacity, or ability to withstand changes. Unfortunately, this is where soilless mixes come up short, particularly for the hobbyist. Large-scale greenhouses can afford the testing equipment that measures soil chemistry, but most homeowners can't, and don't want to bother. I think the easiest way to improve a potting soil's buffer capacity is to add 5% to 10% by volume of loamy garden soil. After using soilless mixes exclusively for several years, I've started including real soil, and it makes a welcome difference in how long the pH stays at an acceptable level. I gather crumbly, compost-enriched soil from my vegetable garden, spread it on newspaper to dry, and then sift it through a ¼-in.-mesh screen. If I didn't have a garden, I'd use the best topsoil I could buy.

Nutrient supply—Whether or not a potting soil contains an initial charge of plant growth nutrients doesn't matter much to me, since it's so easy to meet a plant's requirements with regular doses of water-soluble fertilizer. More important than what a potting soil starts with is its ability to retain nutrients and slowly release

them to the plant. This ability depends on the ingredients in the mix and can't be simply measured, but like buffer capacity it can easily be increased by adding a small amount of loamy garden soil. Ideal potting soil can retain a supply of nutrients for weeks or months, depending on the plant's needs.

Usually it's hard to tell from the bag whether or not a commercial potting soil contains nutrients, and in what proportions and amounts. Adding more uncertainty, some mixes contain high levels of soluble salt compounds that aren't needed as plant nutrients, and may even be damaging to roots. Soil labs have equipment for testing nutrient and soluble-salt levels, but accurate test equipment is too expensive for my budget. Unless I'm fa-

Some potting-soil mixes contain high levels of fertilizers and soluble salts; others don't. This information is rarely specified on the bag. An easy way to reduce uncertainty is to leach out excess elements by running plenty of water through the soil when you pot a plant.

miliar with a mix and know how plants respond to the nutrients it contains, I normally prefer to leach it out and start from scratch. To leach most of the initial nutrients and salts from commercial potting soil, run extra doses of water through the pot at the time you pot a plant (this helps settle the soil in around the roots, too). A few weeks later, start a regular fertilizing program.

Tips for soil shoppers

Potting soil is cheap compared to the plants you buy and the time you spend growing them—it's false economy to let valuable plants languish in inferior soil. If you're disappointed with what you find in a bag, dump it on the garden. At worst, it's an expensive garden-soil amendment;

at best, you'll avoid problems with your potted plants.

If there's a nursery in your area growing plants you admire, ask them what kind of soil they're using, and if they'll sell you some. Whether they mix their own or use ready-mixed, they've probably scouted out the most consistent, economical supply. The nursery staff can answer your questions on the best way to water and fertilize plants in that soil mix. Plants you buy from that nursery will have an easy adjustment if you repot them with the same soil they're used to.

Whether or not you buy through a friendly nursery, I'd recommend purchasing soil that's designed for professional growers. (One way to judge is by the type of information on the label—does it read like it was written by a horticulturist, or on Madison Avenue?) The reason is simple. State and federal regulations don't govern potting-soil quality; market feedback does. Companies that supply commercial growers have in-house testing labs that regularly monitor their product. They have to, because their customers demand it. Nurseries depend on the success of their crops, and can't risk losses due to inferior or inconsistent soil. Companies that market primarily to home gardeners may maintain the same high standards, but they don't get the same kind of feedback. Ask yourself, have you ever complained about a bum bag of potting soil? Rest assured that any nursery that loses a $10,000 crop will make a fuss. Along these lines, I usually buy nationally distributed rather than local brands, because I think the bigger companies invest more in quality control.

It may be hard to find more than one brand of professional-grade soil at any particular garden center. If you want to try different brands, you may have to shop in several places. In remote areas, it may be more convenient to mail-order from greenhouse-supply companies. Catalog prices for major brands look low until you add on the shipping charges; then the cost is about the same whether you buy mail-order or shop locally.

When you find a mix you like, buy it in quantity to get the best price—a 40-lb. bale may cost only three or four times as much as a 4-lb. bag. Share the savings by shopping with a gardening friend, or stockpile the surplus for future use. Kept dry, most mixes can be stored indefinitely. However, don't store bags of wet potting soil, particularly in warm weather—particles of peat and bark begin to decompose, and the soil's pH and fertility are liable to change for the worse. For the same reason, don't buy broken bags that have been stored outdoors in the rain and sun. □

Rita Buchanan is an associate editor at Fine Gardening.

Soil Drainage

What's soggy, what's not, and what to do about it

by Richard E. Bir

When I'm asked to diagnose an ailing plant, the most frequent culprit is soil drainage. The gardener did not take into account the one factor that overrides proper planting, pruning, mulching and fertilizing. If you do everything else right but you plant something that likes perfect drainage in a soggy spot, you're bound to fail.

Plant roots must breathe. They need oxygen to live, and they get it from the air that fills spaces between particles of soil. But in soggy soil, water displaces air. When there's not enough oxygen, roots suffocate. Every plant has a different capacity for dealing with a lack of oxygen; for instance, river birch tolerates temporary flooding, while paper birch does not. But in general, poor drainage kills plants.

Soggy soil also creates conditions that foster some disease-causing fungi. Poor drainage is every bit as much the culprit as the fungi. You may control the disease temporarily, but the trouble will recur as long as the plants have soggy feet.

Before I explain why some soils are soggy (and others are dry), let's agree on a few terms. The lingo of soil drainage describes how long water hangs around after it arrives. It doesn't matter how the water arrives: from the sky, through an expensive irrigation system or out of the end of a garden hose. The term "good drainage" means excess water leaves at an acceptable pace. "Poor drainage" means excess water stays in the soil too long—puddles persist for hours after a shower, or the soil is regularly soggy. "Excessive drainage" means water drains so fast that many plants will need frequent watering.

An easy test to determine drainage

If you're not sure what sort of drainage you have, here's a simple test.

You will need a 46-oz. can, the kind grapefruit juice and tomato juice come in. Remove the top and bottom of the can. Dig a hole 4 in. deep and set the can on the floor of the hole. Firm soil around the can so that water can't slip under the bottom edge. Then fill the can with water to the top. Now wait an hour.

How fast the water disappears depends on drainage. If the water level drops at least 2 in. in one hour, you have roughly normal drainage, adequate for most garden plants. If the level drops more than 5 in. in one hour, you have excessive drainage. You may have to make the soil retain water (by adding plenty of organic matter) or be willing to irrigate many plants. If the water doesn't seem to drain at all, you have poor drainage; just remember that bog and wetland gardens can be beautiful. —R.B.

Filling a partially-buried, bottomless can with water is the first step in testing the rate of drainage in your soil.

One hour later, the water level has fallen 2 in., a sign of adequate drainage. A drop of 5 in. is fast. If the water fails to drop at all, you have a problem.

The causes of poor drainage

Drainage depends on soil structure. If there are lots of big air spaces, drainage is good. If there are few air spaces, and they're mostly small, drainage is poor.

Mineral particles, the largest portion of most soils, play a leading role in drainage. If they're little rocks, the soil is sandy or gravelly, and it usually has excellent drainage because there are lots of relatively large air spaces between the little rocks. I've gardened on sandy soils in Florida and some sandy gravels called soil in New England; they drained so fast I don't think the rain really stopped falling when it hit the ground—it just briefly slowed until it reached the water table or bedrock. When drainage is too quick, the only remedy is to amend the soil with plenty of organic matter: compost, shredded leaves, peat moss and the like.

If a soil has tiny mineral particles, it's usually clayey or silty and tends to pack tightly, leaving little room for water to pass through. Gardeners who garden on clay, including those of us who garden on the red clay of the Carolinas, have several options for coping with poor drainage. I'll tell you about them in a minute or two. In clay soils, a brief shower can create a puddle that lingers for days.

Drainage also varies because of compaction and barriers to water movement. Pathways almost always have poor drainage because our repeated footsteps pack the soil, decreasing or nearly eliminating air spaces. The same thing happens in areas where water repeatedly falls or flows—for instance, under eaves, under the edges of decks and below clogged gutters. Barriers are obstructions, such as a layer of clay, a rock, or construction debris beneath the soil or at the lower end of a garden. Water still flows downhill (gravity works!), and the soil may drain well, but when the water hits the impenetrable barrier, it stops flowing and soaks the soil. Correcting subsurface drainage problems caused by compaction or barriers is seldom easy and frequently involves both heavy equipment and major expenditures. But you may need only to unclog natural drainage, so before giving up, consult a local professional.

Unless you plan to move, you have only two choices when it comes to gardening and poor drainage. Either you accept the drainage you have and choose plants which tolerate or even thrive with poor drainage, or you change the drainage. Most successful gardeners do a little of both.

Woody plants for soggy sites

There are many terrific plants that tolerate wet feet rather nicely (see the list at right). And you don't have to travel far to see examples. Near an upland New England swamp you'll find red maples with pussy willows, alders and deciduous hollies actually in the edge of the water. (Where I was raised, the red maples were called swamp maples, but that was before beautiful selections with names like 'October Glory' and 'Red Sunset' came onto the market.)

A bounty of white flowers signals success on a wet site. The shrub is 'Henry's Garnet', a Virginia sweetspire. It grows well in most soils, but is also adapted to poor drainage. Many other plants can help you garden on wet sites (see the list below).

Native shrubs and trees for moist soils

All plants on this list will tolerate moist soils, and some will tolerate or even thrive with their roots regularly flooded. Consider planting them when amending poor drainage is impractical.

Bald cypress (*Taxodium distichum*)

Bayberry (*Myrica pensylvanica*)

Black gum (*Nyssa sylvatica*)

Button bush (*Cephalanthus occidentalis*)

Carolina buckthorn (*Rhamnus caroliniana*)

Catalpa (*Catalpa bignonioides*)

Cinnamon clethra (*Clethra acuminata*)

Doghobble (*Leucothoe fontanesiana*)

Dwarf azalea (*Rhododendron atlanticum*)

Elderberry (*Sambucus canandensis*)

Flowering raspberry (*Rubus odoratus*)

Hardhack (*Spiraea tomentosa*)

Hearts-a-bustin' (*Euonymus americana*)

Highbush blueberry (*Vaccinium corymbosum*)

Ironwood (*Ostrya virginiana*

Live oak (*Quercus virginiana*)

Mountain spicebush (*Lindera benzoin*)

Ninebark (*Physocarpus opulifolius*)

Pink-shell azalea (*Rhododendron vaseyi*)

Poinsettia tree (*Pinckneyea bracteata*)

Pond cypress (*Taxodium ascendens*)

Pussy willow (*Salix discolor*)

Red buckeye (*Aesculus pavia*)

Red chokeberry (*Aronia arbutifolia*)

Red maple (*Acer rubrum*)

River birch (*Betula nigra*)

Salt marsh elder (*Baccharis halimifolia*)

Serviceberry (*Amelanchier canadensis*)

Silky or swamp dogwood (*Cornus amomum*)

Smooth or white sweet azalea (*Rhododendron arborescens*)

Summer magnolia (*Magnolia grandiflora*)

Summersweet (*Clethra alnifolia*)

Swamp azalea (*Rhododendron viscosum*)

Sweet bay (*Magnolia virginiana*)

Sweet gum (*Liquidambar styraciflua*)

Sweetspire (*Itea virginica*)

Sycamore (*Platanus occidentalis*)

Titi (*Cyrilla racemiflora*)

Virginia or ditch rose (*Rosa virginiana*)

Washington hawthorne (*Crataegus phaenopyrum*)

Water oak (*Quercus nigra*)

Winterberry holly (*Ilex verticillata*)

Witch alder (*Fothergilla gardenii*)

Witherod viburnum (*Viburnum cassinoides*)

Yaupon holly (*Ilex vomitoria*)

All along interstate highways there are plants that tolerate poor drainage. The trees include sweet gums, sycamores and red maples; in the South, you'll also see live oaks and magnolias. Few plants tolerate soggy soil as well as bald cypress or pond cypress. All of these trees also perform beautifully on soils of average drainage. As a bonus, they're hardier than you might think. There are evergreen magnolias that tolerate -20°F and bald cypresses that have grown in southern New England for decades.

Among shrubs, a few native azaleas tolerate wet feet, including the pink-shell (*Rhododendron vaseyi*) and sweetly fragrant natural hybrids of dwarf azalea (*R. atlanticum*) and pinxterbloom (*R. periclymenoides*) from along the Choptank River on the border between Maryland and Delaware. My favorite Choptank hybrid is 'MaryDel', which has soft pink flowers with a spicy fragrance. When mixed with powder blue forget-me-nots and scarlet cardinal flowers (*Lobelia cardinalis*), these azaleas make me long for more spots with limited drainage.

One of the bonuses of moist-soil plants is that so many seem to attract a variety of wildlife. Butterflies are always hovering around Joe-Pye weed and New York ironweed in wetlands in late summer. Deciduous hollies north to New England and the evergreen yaupon hollies of the South feed and provide habitat for birds in the winter, while elderberries, highbush blueberries and many viburnums and serviceberries feed wildlife throughout the summer and fall.

The list of plants adapted to moist soils is long. You're limited only by your willingness to search for nurseries that grow them.

Improving poor drainage

Most of us want at least some plants that don't do well in soggy soils. To grow them, you have to improve poor drainage by amending the soil or channeling water. Both methods work, even if you live where water fills any hole you dig and your soil might be used to make pottery. Start by studying your yard during or just after a rainfall to learn where water flows and where water tends to puddle.

Make raised beds—Because water flows towards low spots, one common way to improve poor drainage is to create raised beds. In short, borrow a little soil from one area and pile it in another. The area that lost the soil is now lower and may become a path for drainage. The higher the raised bed, the drier or better-drained the soil near the top will be. I usually suggest starting with a raised bed at least 6 in. deep. The bed can be just a natural-looking mound, or you can support it with railroad ties, landscape timbers, rock or brick.

One caution: raised beds can modify the way water flows through the garden. If too much water flows too fast, it can carry things along with it and lead to erosion. So be careful how you place a raised bed. On the other hand, in my garden, I've made a bed that combats erosion. I've placed a rock-edged raised bed to direct and slow the torrents of water that rush off my paved driveway on a steep slope. Hostas and ferns planted under 'Snow Queen' oakleaf hydrangea, 'Henry's Garnet' sweetspire (*Itea virginica*), pink-shell azalea and summer-sweet (*Clethra alnifolia*) provide backyard privacy and a feel of woodland coolness, as well as functional erosion control.

If your soil is clay, however, simply building a raised bed won't accomplish much; you'll just end up with a pile of brick clay. But adding air spaces can improve clay soil. I know a beautiful woodland garden built where driveway runoff once flowed. The gardener diverted the runoff and fluffed up the compacted soil in the former water course. Bad soils don't need to stay bad.

Organic matter can help or hurt—Adding organic matter to soil can improve poor drainage or make it worse. Some organic matter, such as peat moss, acts like a sponge, absorbing water and holding it around plant roots, which is exactly what you want if your drainage is a little too good, as it often is in sandy or gravelly soils. But if your drainage is poor, you should add organic matter judiciously. Peat moss may hold too much water. I caution anyone in an area prone to heavy rains to be cautious with peat as a soil amendment, because it can retain more water than is good for many plants. Our research at North Carolina State University has demonstrated that some plants perform better if organic matter with a more limited sponge action, like pine bark, is mixed uniformly with clay or clay loam. The plants are generally those which grow well in humusy woodland soils—azaleas, camellias and mountain laurel. Eventually, organic soil amendments will decompose, but they help plants become established by aerating the soil for the first few post-planting years. If you mulch and plant properly, your plants will continue to thrive.

You might think a little sand would improve the drainage of clay soils. But remember that adobe and other bricks are a mixture of clay and a little sand, and you certainly don't want your bed to be as hard as a brick. So, if you plan to amend your clay with sand, you must do two things. One, make sure the sand is coarse rather than fine, and two, be sure to add enough so that you have more sand than clay in the root zone—that's a lot of sand, more than is practical in many cases.

A planting trick

It's possible to plant in a way that minimizes problems with poor drainage. My advice is to dig a wide hole—five times as wide as the rootball if you can do it—because that circle is where the roots of your plant eventually need to grow. Roots penetrate soft soil more easily than compact soil, and you've softened that soil by digging it. If you are planting into an existing landscape, do the best you can, but don't destroy one valued plant to dig a planting hole big enough for another.

The depth of a planting hole is crucial in soggy soils. Never set plants deeper in your landscape than they were growing in their nursery site. In fact, many successful gardeners plant so that one-third to one-half of the shrub's rootball is above the soil surface, because anything planted high will survive a deluge. However, they do not leave the rootball exposed. They cover it with organic mulch, such as leaves, bark or pine needles, a few inches deep, then watch to make sure the rootball does not dry out for the first few months. New roots need time to reach down into your garden soil, and anything planted high will suffer in a drought until it has become well established. □

Richard E. Bir is a horticulturist at the North Carolina State University Mountain Crops Research and Extension Center in Fletcher, North Carolina. He is the author of Growing and Propagating Showy Native Woody Plants, *the University of North Carolina Press, 1992.*

Composting Leaves
Recycling fall's bounty to improve your soil

by Mark Kane

Composting the fall crop of leaves yields brown gold in spring. Dry, brittle leaves turn into dark, rich humus, a much-decomposed form of organic matter with strongly beneficial effects on the soil. Humus makes soil more open by binding soil particles together, and increases the soil's capacity to hold moisture. Humus also releases plant nutrients gradually for years. When you use leaf compost regularly, you see the difference in tougher, more vigorous plants.

Composting leaves is straightforward. If you choose to let nature do most of the work, you can just pile leaves in an enclosure and come back in a few years to collect the finished compost. By contrast, when you compost yard wastes, hay and the like (as Will Bonsall explained in *FG* #15, pp. 34-37), you must see that every part of the compost pile reaches temperatures around 150°F, hot enough to kill weed seeds and plant diseases. If you choose to hasten leaf composting, you must ensure that the pile always has adequate air and moisture, requirements that oblige you to rebuild the pile periodically and to control its moisture. That's all there is to it. Many gardeners have heard that leaves are balky and will not compost without special ingredients and techniques. Not so. (See *FG* #16, p. 49, for a test of special ingredients and nutrients.)

Recycling leaves is also good ecology. When I see black plastic bags of leaves lined up curbside each fall, waiting for a ride to the municipal dump, I can't help but dream of the day when a compost bin full of leaves will be as common a part of every yard as a lawn.

Composting basics

Composting is a natural process, helped along by the gardener. In nature, a host of soil creatures specialize in breaking down the annual litter of fallen leaves, stalks, stems, twigs and branches. When you heap up leaves to make a compost

Leaves change dramatically during composting. At top are freshly shredded leaves. Kept dry, they'll look like the center pile in six weeks. Kept moist and turned regularly, they'll break down and get to be the uniform dark brown of the finished compost at bottom.

pile, you spread a banquet that attracts and feeds many of these creatures, particularly bacteria and fungi. As long as they are surrounded by air, moisture and food, the microorganisms multiply, and the leaves break down rapidly.

An actively composting pile heats up, as the metabolism of billions of microorganisms releases warmth faster than it can escape the pile. Its coarse texture and multitude of dead air spaces make the pile an effective insulator. At the center of the pile, for short periods, the temperature can reach 160°F. On a frosty morning in fall, you can pull back the top layer of leaves and watch steam rise. I've heard gardeners who are new to composting ask if a compost pile can catch

Chopping a pile of unshredded leaves into 4-in. slices with a mattock breaks up compacted layers and replenishes air for microorganisms that break down the leaves. When author Kane finishes chopping, he'll rebuild the pile.

Photos: top, Mark Trela; all others, staff.

fire. The answer is no. A compost pile is far too moist to ignite at the relatively low temperatures it can achieve. What's more, the heat of a compost pile is self-regulating. As the temperature rises, microorganisms die in droves because the supply of air is dwindling and because they cannot tolerate the heat. At 150°F to 160°F, the remaining population is small and incapable of driving the temperature higher. As heat escapes the pile, the temperature peaks and then falls.

A compost pile that has heated up and cooled off will heat up again only if you stir it to replenish air. This means breaking the pile apart and reassembling it—a practice called "turning." As long as the leaves haven't broken down into humus, the pile will heat up every time you turn it, though in the late stages of composting, when the remaining nourishment in the leaves is harder for microorganisms to extract, the temperature rise will be slight.

You have to control moisture in a compost pile. Too little water makes the pile a desert. As you heap up leaves, water them copiously. A pile is properly moist when it feels like a slightly damp sponge. A sodden pile is too wet. It will settle dramatically under its own weight, driving out air, while the excess water displaces more air, and the usual air-breathing microorganisms will give way to their anaerobic kin, microorganisms that don't require air. The pace of composting will slow and the pile will acquire an unpleasant, swampy smell. Happily, you can easily correct overwatering. To restore active composting, just break the pile apart, allow the leaves to dry a bit, and then rebuild the pile.

Compost bins
When you assemble leaves in a compost pile, it helps to have an enclosure—a compost bin—to keep them from tumbling and sprawling and blowing away on the wind.

I've built several enclosures over the years, including substantial wooden cribs, and now favor the simplest and most easily disassembled bin of all—a section of fence wire drawn up in a circle. I use 42-in.-high wire, but 36-in. or 48-in. will do fine. I cut the section roughly 12 ft. long, with a few horizontal wires bent in hooks at one end to close the bin in a circle 4 ft. in diameter. Small as that may sound, you need a lot of leaves to fill the bin—the equivalent of 12 large yard bags.

When I start a pile, I leave the fence wire aside at first. I haul two or three loads of leaves and heap them in one place until the pile looks about the right diameter. Then I pull the wire around the leaves and hook it shut. Leaving the fencing aside at first saves me from hoisting a lot of leaves 42 in. high to clear the wire. I

A 2-ft. thermometer records a temperature of 110°F in early spring, as a compost pile awakens from its winter sleep. With two or three turnings, the pile will yield finished compost.

Easy leaf collecting
If you want to save leaves for composting, I highly recommend raking with help from a plastic tarp. In the same time it takes to fill a bag with leaves (scooping up the leaves by the armload and stuffing them awkwardly into a collapsing plastic sack), you can rake five bags' worth of leaves onto a tarp. Pulling a loaded tarp, even one laden with wet leaves, is easy. Plastic slides over grass as if it were greased.

My tarp is a piece of 4-mil polyethylene—not only inexpensive but fairly long-lived. I bought a 10-ft. by 50-ft. roll for $6, and cut a 12-ft. length for the tarp, leaving several replacements on the roll. If I confined hauling to the lawn area, the plastic would last indefinitely, but every year I pull a load or two over the driveway or a tree sprout clipped short by the lawn mower, and the plastic tears. Nonetheless, it took my first tarp four years to spring a big leak. I retired it to a career as a painting dropcloth, and cut a new tarp from the original roll. If you know your tarp will see rough duty—repeated trips over rocks, concrete, asphalt, gravel or woody stubble—I suggest you buy a woven or reinforced tarp.

I discovered early on that, on slopes, a load of leaves tends to slide off polyethylene. My remedy may not be elegant, but it works. I gather up two corners and the edge between them and cinch them with twine into a pigtail. At the cinched end, the tarp curves into a scoop that holds the leaves.

I spread the tarp beside a leaf pile and rake the leaves onto it, pushing them into the cinched end; to keep the tarp in place, I hold down the edge with a board. When the tarp is full, I toss the board on the leaves, gather up the free end of the plastic sheet and drag the load to the leaf pile. To dump the leaves, I grab the scoop end of the tarp and pull it over the other end. Within ten seconds, I'm on my way back for more leaves. —M.K.

can get about half a pile heaped up before I have to fence it in.

I use the same gambit when I turn a pile. I unhook the wire and set it aside. Then I break up the pile, and toss the leaves nearby, heaping up as much of the new pile as I can before I circle it with the wire.

You can place a compost pile almost anywhere you like, but bear in mind that compost attracts roots. One year, I made a compost pile behind a garage overshadowed by trees. I used half the compost the next spring and came back in the fall to take the other half. It wouldn't budge. The pile was bound together by roots, many as thick as clothesline. These days, before I build a leaf pile near trees or shrubs, I first lay a 4-ft. square sheet of plastic on the ground. If you make more than one compost pile, take care not to group them where the plastic may suffocate enough roots to harm a tree or shrub.

Comparing options
Before you plunge into leaf composting, consider how hard you're able and willing to work at turning your leaf pile. You have three options: you can use shredded leaves, which are fairly easy to turn; you can use unshredded leaves, which take hard work to turn; or you can skip turning altogether.

Unshredded leaves flatten and knit together in dense layers that resist turning. A spade is useless. I used to turn leaf piles with a pitchfork, working from the side and easing the tines between layers, but it was arduous work. You'll find yourself hoisting heavy clumps of leaves, and before you can toss them nearby to start the pile over, you'll have to shake and tease the clumps apart. I now prefer a friend's method—chopping through the layers with a mattock. I work in from the edge, 4 in. at a whack. I can chop a good-size pile in ten minutes of hard work, and heap the scattered leaves into a new pile in ten minutes more. No matter how thoroughly I break up or chop unshredded leaves, the finished compost always has clumps that have remained intact and largely escaped composting, but I don't mind. The clumps are ill-suited to spading into the soil or to mulching small plants, but fine for mulching shrubs and young trees.

Shredding leaves before you compost them has several attractions. Shredded leaves are far easier to mix and turn than unshredded leaves, and the finished compost has a uniformly fine texture that is well suited to mulching or to tilling into the soil. Also, shredded leaves take up less space in the early stages of composting, so you can pile more in the bin.

I've made compost from shredded leaves and from unshredded leaves, and on the whole, I prefer unshredded leaves for two reasons—I enjoy hard work, and I don't own a shredder. If I had to guess, I'd say that there's no more work all told in turning unshredded leaves than there is in the two-part process of sending leaves through a shredder and turning a pile of chopped-up leaves.

You can also choose to eliminate turning the pile altogether and make what is usually called "cold" compost. Once a newly-made pile heats up and cools down, microorganisms that prefer moderate temperatures move in. Since the pile settles and is short on air, they work at a slow pace. They're soon joined by earthworms, which churn through the pile, leave nutrient-rich castings, and multiply prodigiously. After two to four winters, you have finished compost.

Composting without turning is an attractive prospect, but there are trade-offs. You have to wait several years for finished compost. Also, if you build a new pile each fall, as you should, you'll need room on your property to keep three or four piles going year in and year out. Finally, you have to cover the piles with long-lived tarps to keep out heavy rains and melting snow, and water the piles promptly when they become too dry.

I choose to hasten composting in order to have finished compost by spring. I turn the pile each time it cools down, roughly every week. Most years I have half-finished compost when cold weather puts the pile to sleep for the winter. Then, to fend off snow and spring rains, I cover the pile with a square of 4-mil plastic. I remove the wire cage, shape the top of the pile into a peak, drape the plastic over it, and weight the plastic with a rock or

Covering a leaf pile for the winter with a sheet of 4-mil black polyethylene, weighted with a rock tied in each corner, keeps snow and rain from leaching away nutrients.

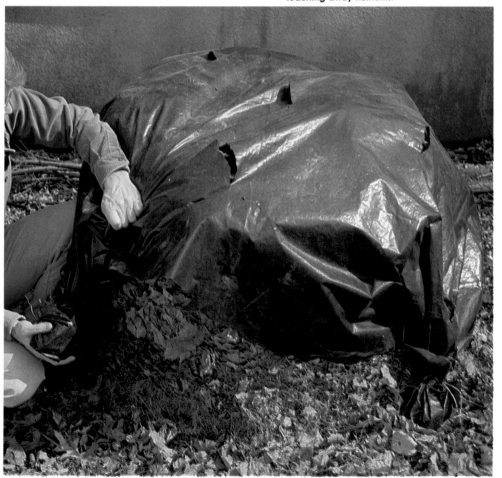

brick tied in each corner to keep the wind from carrying it off. When the ground is ready to work in spring, I start turning the pile again, and usually have finished compost after two or three turns.

Using leaf compost

When is leaf compost finished? Composted leaves develop the insubstantial texture of wet newsprint and a uniform dark color—what were once dull-brown leathery oak leaves and red papery maple leaves are now indistinguishable. If you grab a handful at random from the pile, it breaks readily into fragments. The pile has a pleasant smell that reminds many people of a forest floor.

You can use a leaf pile before it becomes finished compost. Some springs, I lift layers intact from the pile and place them in overlapping patterns for mulch. Few weeds can push through a sheaf of matted leaves. By fall, the leaves have broken down nicely, thanks mainly to earthworms, which burrow in the seams.

I use finished leaf compost several ways. The texture and soil-building properties of finished compost make it especially well suited for enriching the soil when you break new ground, for maintaining soil fertility in the vegetable garden, and for mulching. Because finished compost is soft and crumbly, I can spread an inch or two over the vegetable garden and spade it in uniformly with very little effort. Applied each spring, the compost can dramatically increase the tilth and moisture-holding capacity of the soil in a few years. Each application also releases nutrients slowly for years.

Leaf compost makes wonderful mulch. It spreads nicely and presents a smooth, uniform surface. I rarely have enough leaf compost for my entire garden, so I reserve it for small, densely-packed plants or plants in prominent locations, where appearance counts. The crumbly consistency of leaf compost suits it for delicate mulching. I can rub it between my palms over a small-leaved ground cover such as vinca and let fine pieces sift between the plants. If you try that with uncomposted leaves, you smother the ground cover.

Making leaf compost is rewarding work. It's good for the garden, and good for the soul—you are recycling energy and nutrients, as nature intended, and sparing your local dump. Leaf compost also gives you a new perspective on trees. They cease to be shade-casting problems that make it hard to garden in parts of your yard. They become allies instead, generous friends who make a gift of their hard-won bounty to enrich your garden. □

Mark Kane is an associate editor at Fine Gardening.

Get Started in Composting

Turn yard waste into a valuable soil booster

The four elements of compost: Brown materials (leaves), green materials (grass), water and air.

by Steven Cline

Long before it became a '90s buzzword, composting was something we just did. On my family's farm we had to make the most efficient use of all our resources. The vegetable scraps, the rabbit manure and the animal bedding all went into the pile, there to decay naturally into a fine, dark, moist and crumbly substance.

We did not have any rules for how high it could be piled or how much we could add of one thing versus another. Compost just happened, and when it got to the stage where grapefruit rinds and eggshells were no longer recognizable, it was spread out on the flower beds and the two-acre vegetable garden and tilled in. It improved the structure, water-holding capacity and aeration of the soil in which we grew our food. It buffered the garden from freezing and thawing, slowed erosion and supplied our crops with some nutrients. Not bad for a heap of "garbage."

Composting was easy for us, and it can be for you, too. It's not rocket science—just a natural process you can coax to make it happen faster. Even if you set it up without the perfect ingredients or conditions, it's hard to fail.

Starting off

Before you begin composting, you'll need to create a work space. You'll want it to be close to where you will use the finished compost and convenient to the raw materials. An ideal site is no more than a hose-length from a water faucet, is out of the wind, is not under trees or within reach of their root systems and has good drainage so the bottom of the pile will not get soggy. It can be in either sun or shade. You'll also probably want it concealed from sight somewhat, as you would any work area.

The amount of space you'll need depends on how much "waste" your garden produces. I use about 30 sq. ft. for a 7,500-sq.-ft. (0.2 acre) yard. My lot is a fairly typical size for an urban area, and about half of it is either lawn or garden. As a rule, allow between 60 and 90 sq. ft. of composting work space for a lot up to a half acre in size.

Green materials: Kitchen scraps and green yard waste are nitrogen-rich materials for your compost pile.

I would recommend construction or purchase of a composting bin. The main purpose of a bin is simply to keep the materials contained, so, it can be almost anything that will hold a minimum of one cubic yard (3 ft. tall, wide and deep) of materials. I've found that the simplest and most effective "bin" is a 12-ft. section of galvanized dog fencing (having rectangular 2-in. by 4-in. openings) formed into a circle. Get the heaviest gauge wire you can find. The cost should be less than $10, and you can find fencing at any well-stocked hardware store.

Ingredients

Compost ingredients come in two forms: brown (carbon-rich) and green (nitrogen-rich). "Brown" materials include dried leaves, straw, newspaper, wood chips, sawdust and even clothes dryer lint. Most of them are actually brown in color and dry to the touch. "Green" materials include green yard debris, grass clippings, manure, fruit and vegetable peels and rinds, coffee grounds and even hair. Many of them are actually green in color and usually have a high moisture content.

The idea with browns and greens is to mix the two together in proportions that provide a balance of carbon and nitrogen. An adequate supply of each is needed by the microbes responsible for decomposition. These microbes

Brown materials: Newspaper, leaves and dry yard waste provide carbon for the microbes that create compost.

Tips for building a successful compost pile

- ♦ *Lightly wet down each layer as needed to make it about as moist as a damp sponge; green materials may not need any additional moisture.*
- ♦ *Alternate layers of brown and green materials, making each layer no more than a few inches thick.*
- ♦ *Bury kitchen scraps in the bottom or center of the pile where animals won't dig for them.*

occur naturally in the air and soil. To provide them with the balance of foods they require, mix one part green materials to two parts brown materials, measured by volume, not weight. In other words, for each shovelful of greens, you'll need two shovelsful of browns.

Building the pile

To build a compost pile, assemble the correct proportions of green and brown materials and begin layering them alternately in your bin according to the proportions given above. Wet each layer down with a hose to the point where the materials just glisten or where they become sticky when you put your hand into the pile. Make each layer no more than a few inches thick, and put kitchen scraps at the bottom or center to avoid the problem of animals foraging for food and disturbing the pile.

At certain times of the year, you may not have enough of either browns or greens to build a pile in the "correct" proportions. In the fall, for instance, dead leaves are abundant, but green materials are scarce. One solution is to stockpile brown materials until some greens are available in the spring and then build the pile. I have used this method, piling my leaves in a 6 ft. x 6 ft. crib for the winter. However, you can compost brown materials with little in the way of greens. Make the pile as you normally would, but between the layers of leaves sprinkle a little topsoil. The topsoil settles the pile down, adds more microbes to the mix and helps retain moisture. The pile will decompose, but it will take longer.

In spring and summer, you can find yourself with an overabundance of greens—usually grass clippings. Spread clippings thinly (1 in. thick) on garden beds as a mulch; they'll quickly decompose. Or practice "grasscycling" by leaving clippings on your lawn.

The process

A compost pile "at work" is fascinating. Microbes begin feeding on the wealth of food immediately, causing a noticeable rise in temperature at the center of the pile. After two or three days (if that long), the center of the pile will feel hot to the touch. Temperatures can reach 160°F. Between 140°F and 160°F, weed seeds, insect larvae and even some microbes will begin to die off. The pile will shrink to perhaps 40% of its original volume.

The balance of nutrients, air and/or water changes as decomposition progresses, which in turn stops the heating process. After seven to 14 days, the temperature will drop down to below 100°F. For fast decomposition, you will need to turn the pile by mixing the contents with a manure fork or pitchfork. You literally stick the fork in and fluff up the compost-to-be or turn the whole pile over. Turning a pile means work, and I like to leave it optional, depending upon whether you want the process to go faster or slower. If you turn it, you'll get compost faster, but in either case, compost will happen.

Turning should result in a second heating cycle within a few days. The

RESOURCES

Your state's Cooperative Extension Service has information on composting at little or no charge. The author also recommends the following:

Composting to Reduce the Wastestream. Northeast Regional Agricultural Engineering Service, Cooperative Extension, 152 Riley-Robb Hall, Ithaca, NY 14853; 607-255-7654. booklet, $7; 44 pp. Item # NRAES-43.

Let It Rot! by Stu Campbell. Storey Communications, Inc., Schoolhouse Road, Pownal, VT 05261; 800-441-5700. $8.95 plus $3.25 shipping and handling, paperback; 132 pp.

Backyard Composting: Your Complete Guide to Recycling Yard Clippings. BookMasters, 1444 State Rte. 42, RD 11, Mansfield, OH 44903; 800-267-4391. $6.95 plus $3 shipping; OH residents add sales tax; paperback; 96 pp.

Backyard Composting, Missouri Botanical Garden, Kemper Center for Home Gardening, P.O. Box 299, St. Louis, MO 63166; 314-577-9440. $12.50, 26 min. video.

Illustration: Michael Rothman

temperature will rise again and may get as high as it did the first time. By the end of this heating cycle, the pile will have shrunk so that high temperatures are hard to attain—small piles lack the mass to prevent heat loss. Each time you turn the compost, it will heat up more slowly and to a lesser degree. Continue turning the pile every so often until it no longer heats up at all, or let it sit undisturbed until you can see fine brown or black soil-like particles at the bottom of the bin. At that point, you have finished compost to use in any number of ways.

Finished compost is completely safe to touch and not at all yucky. It looks and smells a lot like soil. It makes an excellent amendment for your garden beds. You can sift out the finer particles using a ¼-in. screen to get an excellent potting soil additive. Toss the coarse, leftover compost into the next pile to continue breaking down, or use it as a surface mulch.

Fine-tuning

If you feel like going the extra mile and really speeding up the process, try shredding the materials. Smaller particles have an increased surface area for microbes to work on, so the process proceeds much more quickly and generates more heat, which in turn accelerates the process even more. A chipper/shredder is useful to shred garden debris. A lawn mower is effective for leaves and nonwoody waste. Newspaper can be shredded by hand, and kitchen scraps can be cut into small pieces.

To further accelerate decomposition, you can try adding what's called "compost starter," a microbe-packed powder sold nearly anywhere you find compost bins and tools. I have yet to see scientific proof that starters are effective in small-scale composting, but some gardeners have apparently had good results, and you may wish to experiment.

Problems

There are really only two problems you might encounter: a pile that smells or a pile that doesn't heat up. Don't despair. Just review the basics: Are the proportions of green and brown materials right? Is there too much or too little water or air?

Finished compost: Your plants will love it as a potting soil ingredient (after sifting), mulch or soil amendment.

A pile that is unusually odorous can be caused by an excess of green materials, an unmixed deposit of grass clippings or other green materials, or an excessively wet pile. Make sure the mix of brown and green materials is right and that there are no big clumps of greens. Place the pile in a spot that drains adequately so water does not accumulate around the base, and if it does get too wet, fluff it up to bring air in.

A pile that doesn't heat up is most likely due to brown materials that are too coarse, a pile that is too small, a total lack of water or a lack of nitrogen. To remedy the problem, make sure you have more than just, say, twigs for brown materials, and that the pile is at least one cubic yard in size. Rebuild the pile, following the instructions I've given for wetting the pile and for mixing browns and greens, and your pile should start cooking.

Also take into account the outside air temperature. When it's below 40°F, expect less activity. Assuming everything else remains in order, once temperatures rise, the process of decay will take off.

Tools and gadgets

One of the real attractions of backyard composting is the fact that setup is cheap and recurring costs are small or nonexistent. I've tried many approaches and have but one conclusion: The best setup is the simplest. Spend your money on good tools instead of intergalactic bins. Buy a good manure fork and lightweight scoop shovel (a shovel with a deeply curved blade to hold loose materials) to turn and "harvest" the pile. A pitchfork, a hose-end sprinkler with a shut-off control (to make wetting down the layers easy without wasting water) and a machete for chopping are also useful.

Last but not least, buy a long-necked composting thermometer with a good dial for easy reading. That way, for every adjustment made to the pile, in 24 to 48 hours the gauge will indicate if your efforts are a success. Watching your pile hit 120°F when it's cold outside is guaranteed to charge your composting spirit! □

Dr. Steven Cline is manager of the Missouri Botanical Garden's William T. Kemper Center for Home Gardening.

Making Hot Compost

Transform yard wastes into a valuable soil amendment

by Will Bonsall

Compost is the backbone of the soil-fertility program on our farm. My wife, Molly, and I grow much of our food and try to live as self-reliantly as possible. We make compost from free materials we have on hand, such as tree leaves, crop residues, hay, wood ashes and kitchen scraps. For our efforts, we get a complete fertilizer—not just a soil conditioner or a bacterial starter or a trace-mineral source—and we seldom spend a penny for soil amendments. Just as important, we're recycling all the organic material we generate.

Our composting methods work for gardening on almost any scale. We make large quantities of compost, enough to spread about ⅛ in. deep over our one-acre garden each year. (We'd spread it deeper if we had more.) That translates into about 16 tons of compost, but don't let that daunt you. You can easily make smaller amounts. All you have to do is scale down the operation. For example, to cover a 4,000-sq.-ft. garden with ⅛ in. of compost, you'll need about 1½ tons of it, an amount that can be made in a 4-ft. by 4-ft. by 4-ft. bin.

The raw materials for making compost are readily available everywhere. We don't use any animal products, such as manure, bone or blood meal, or feathers, in our compost. Instead, we rely exclusively on plant residues to build our "veganic" compost. Even gardeners who don't share our vegetarian principles might find that our compost ingredients suit their purposes. Animal manures are often tough to come by for city and suburban gardeners, and neighbors may have concerns about possible odors.

If you want to follow our lead, you don't have to own acres of forest, pasture and farm to collect an abundant supply of compostable vegetable matter. Most urban and suburban yards are filled with leaves. If you can't find enough in your yard and in your neighbors' yards, check

Bonsall makes hot compost. During the process, temperatures in the pile get high enough to kill insect larvae, pathogens and weed seeds. To get the compost cooking, Bonsall builds each new pile all at once, using layers of assorted materials he's stockpiled. Here, he forks a load of hay into a new pile.

your local dump. Many town dumps stockpile leaves, which are frequently free for the hauling or go for a minimal charge. Weedy hay often grows abundantly in vacant lots, and it's usually quite easy to get permission to cut it. Later on, I'll mention other readily available compostable stuff, both vegetable and animal.

Why hot composting is best

If you've been gardening for a while, you've probably heard as many different recipes for compost as there are for vegetable soup. I think hot composting, in which the piles reach 160°F, is best. Heat begins to break down plant fibers. More important, it destroys insect eggs and most disease spores and weed seeds in the crop residues. Let me emphasize this: If you use compost that hasn't been heated through and through, you may be strewing your garden with all kinds of pests, diseases and weeds, which can more than offset the benefits of the added fertilizer. For the same reason, hot composting is far better than just tilling in crop residues.

Getting a pile to heat up is simple, as long as a few key requirements are met. (For the details, see "How composting works" on p. 39.) The pile must have a certain critical mass, a proper ratio of high-nitrogen materials to high-carbon ones, and plenty of air and moisture. We also like to include a mineral source such as soil, wood ashes or lime, though it's not essential.

A 4-ft. by 4-ft. by 4-ft. pile should be enough mass to get things cooking. Build the pile as quickly as possible—in a few days, rather than over a period of several months. If stuff is tossed on the pile a little at a time, it will eventually rot, but it won't build up enough heat to do any good. Building the pile at one time also allows us to blend about the same amount of each ingredient each time, which results in fairly uniform batches of compost. In order to have a variety of material on hand, we stockpile our organic material as we collect it. Otherwise, some piles would be nearly all leaves, some mostly hay and others just kitchen scraps.

We store our leaves and hay in a holding crib 8 ft. square. If you'd rather not bother with constructing anything like this, you can keep the material in wire cages or under tarps.

During the winter, we store our kitchen garbage in barrels. If you live in a more populated area than we do, and are concerned about flies and smells accumulating around the kitchen scraps, layer them with sawdust in plastic garbage cans with tight-fitting lids until you're ready to build the pile. (Sawdust is often available free from local lumberyards or woodshops.) We don't have meat or fat scraps, but if you do, you might want to exclude them from your compost, since they're slower to decompose.

The proper ratio of nitrogen-rich and carbon-rich material helps generate heat. Nitrogen is the spark that gets the whole thing going. You don't need as much of it as you do of carbon, but without nitrogen the pile will decompose very slowly. We rely on freshly cut cover crops (such as red clover), fresh green hay and kitchen scraps. Fresh grass clippings, green weeds and animal manures are high in nitrogen, too. (Don't use cat manure or cat litter—they can transmit toxoplasmosis, a disease that can cause blindness.) Leaves and dried hay are good carbon sources, as are sawdust and dried grass or weeds. The recommended ratio is between 15 and 40 parts high-carbon material to each part of nitrogenous material. With such a wide range, you don't need to worry about precisely computing the proportions, a difficult task since many materials, such as cured hay, are high in both carbon and nitrogen. You'll soon learn to judge whether you've got it right by the results. If the pile doesn't heat up, even though it's wet enough, it's probably too low in nitrogen. Piles that smell strongly of ammonia are too high in nitrogen. In between is a wide margin for error.

Microorganisms that break down organic matter need water and air to do their job. Too much water suffocates them; too little decreases their activity. A combination of rainfall and hand-watering keeps our piles plenty wet. Layering materials of different textures as the pile is built and then turning the pile from time to time provides plenty of air.

I think that it's important to include minerals in the compost, though they're not crucial. We add soil as a source of minerals, to retain moisture and to help pack down the other materials. We also add wood ash, which we generate by the barrelful, in small quantities, about 10 lb.

per ton of compost. This is enough to have an effect without making the compost too alkaline.

Building the piles

Our system consists of the holding crib and seven 8-ft.-long by 5-ft.-wide by 3-ft.-high wooden bins, set side by side directly on the soil. I'd recommend a minimum of two bins, so you can toss the compost back and forth between them, but they can be much smaller than ours. (See the drawing and directions for making a small bin on p. 38.) The boards between each of our bins are removable, making it easy to turn stuff from one to the next in line. The length we've chosen isn't critical, but it makes it easy to tell at a glance how much compost we've made. An 8-ft. by 5-ft. by 1-ft. pile contains 40 cu. ft. of compost. Finished compost weighs close to 50 lb. per cubic foot, so for every foot of depth we have about one ton of compost. In our current setup, the first three bins are open to the sky. The next four have sloped roofs with a slight space between the boards, which allows some rain to leak through onto the piles without getting them too wet. In retrospect, I've concluded that it would be better to cover all the bins.

To build the pile in the first bin, I alternate layers of different types of materials: high carbon and high nitrogen, wet and dry, mineral and vegetable, and coarse and fine. To keep everything from wadding up, especially the tree leaves, I make thin layers of each material. Typically, I put down about two to three bales of hay, then roughly

Bonsall recycles free organic materials, such as leaves, hay and kitchen scraps, into compost, a rich soil amendment and fertilizer for the farm where he grows much of his own food. Here, Bonsall displays partially decomposed organic material in one hand, and finished compost in the other.

Building a small compost bin

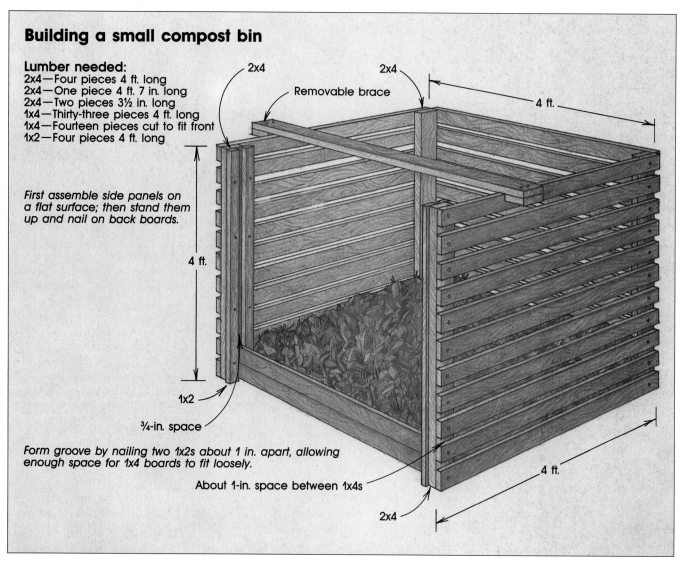

Lumber needed:
2x4—Four pieces 4 ft. long
2x4—One piece 4 ft. 7 in. long
2x4—Two pieces 3½ in. long
1x4—Thirty-three pieces 4 ft. long
1x4—Fourteen pieces cut to fit front
1x2—Four pieces 4 ft. long

2x4

2x4

Removable brace

4 ft.

First assemble side panels on a flat surface; then stand them up and nail on back boards.

4 ft.

1x2

¾-in. space

Form groove by nailing two 1x2s about 1 in. apart, allowing enough space for 1x4 boards to fit loosely.

About 1-in. space between 1x4s

2x4

4 ft.

A series of bins contains the compost, which is turned from one to another as it decomposes. Turning the pile adds air, which is essential for the microorganisms that break down the organic material. Here, Bonsall inspects the compost to decide whether it's ready to spread in the garden.

two bushels of compressed leaves. On top of that goes a wheelbarrowful of garden trash or weeds. I repeat this "sandwich" at least once. Then I add half a wheelbarrow load of fresh clover, kitchen scraps or cider pomice. Occasionally I come by a load of fresh seaweed, which I substitute for part of the clover or garbage. This way, any of the rich nitrogenous stuff that leaches down will be absorbed by the dry bottom layers. On top of this, I continue alternating different types of material, until the pile reaches 5 ft. high.

After every four to six layers of hay and leaves, I add some soil. If I've used a lot of weeds with soil-covered roots, I may not add more than a sprinkling of soil. Otherwise, I add a wheelbarrowful of soil that I've screened through ¼-in. mesh to give it a finer texture.

As I'm building the pile, I climb onto it from time to time and trod it down a bit, especially around the edges. This keeps the pile from being too fluffy and dry. I've heard objections that tromping on the pile pushes out too much air, but

Illustration: Laura B. Goodwin

that's only a problem in piles that include manure or a lot of heavy sod or muck soil.

I cap off the pile with a layer of hay and ½ in. of soil, and then I soak it down with about 150 gal. of pond water. Believe it or not, this doesn't saturate the pile; it just gets it wet enough without allowing water to soak out the bottom of the pile. When I turn the pile later on, I usually add more water.

As soon as the pile is capped and watered, if not before, a tremendous amount of heat is generated from its center. Within a day, I can't poke my hand in more than a few inches without getting burned. After three or four days, the temperature climbs to 162°F throughout most of the pile, with the exception of a few inches on the top and sides. About 18 days later, the pile cools down to a bearable level, near 100°F.

Now it's time to turn the entire wet, steaming pile into the next bin, one forkful at a time. This aerates the pile, blends the ingredients and helps break down fibrous material. It normally takes me 15 to 20 minutes to turn a ton; each of our bins holds four to five tons. If you stood near the pile as it was being turned, you wouldn't believe that this is vegetarian compost. It has the same aroma as fresh cow manure, even though there isn't a cow pie in the heap.

Once turned, the pile heats up again, though it doesn't get as hot as it did initially. After a week, the pile is reduced in volume to half its original bulk. From then on, it's relatively stable and it doesn't heat up again. As the compost "cures" over the next three to four months, remaining coarse particles get smaller, until all that remains is a fine, finished compost that can be tilled right into the garden. After the first couple of weeks, if you want to speed things up you can turn the pile as often as every two weeks during the curing period.

Each time I turn the pile, I check whether it's wet enough. The material should be like a damp sponge—thoroughly moistened but not soggy. If the pile needs water, I soak it with 50 gal. to 100 gal., enough to wet it but not so much that water runs out from the bottom. If the pile is extremely dry, it might be easier to add water as you turn it.

How much does making compost cost us? Our only expense is for labor. We've carefully tracked the amount of time it takes to perform every step, and we've calculated that it takes us nine to ten hours to produce each ton of finished compost. To my way of thinking, that's a very small investment for such a rich reward. ☐

Will and Molly Bonsall farm in Industry, Maine.

HOW COMPOSTING WORKS
by Reginaldo Saraceno

Composting is a process that speeds up the natural degradation of organic wastes by microorganisms. Traditionally, this involves collecting a pile of wastes large enough to be self-insulating—which maintains the high temperatures that destroy pathogens (disease-causing organisms)—and diverse enough to provide adequate nutrients and moisture for beneficial bacteria and fungi. Whether you start with leaves, grass, animal manures, food wastes, wood chips or other organic material, the end result of composting is humuslike material. Used as a soil amendment, compost recycles nutrients, increases moisture retention, and improves water and air movement in soils lacking these qualities.

The primary factors controlling composting are microorganisms, temperature, carbon-to-nitrogen ratio, particle size, aeration and moisture. Understanding how these factors influence composting allows you to build and manage a compost pile, while optimizing the breakdown process.

Decomposition by microorganisms—The bacteria and fungi normally present in organic wastes convert them to compost, producing carbon dioxide, water and heat at the same time. Different microbes function at different temperatures. Microbes that flourish at 50° to 105°F (called mesophilic microbes) start the decomposition process. They consume the most easily available starches and proteins and break them down into simple sugars, organic acids, peptides, carbon dioxide and water. As the mesophilic microbes increase their metabolic activity and their numbers multiply, the compost pile heats up. Above 115°F, microbes that thrive at higher temperatures (called thermophilic microbes) increase in number. The temperature of the pile then rises even higher, and at the same time the population of mesophilic microbes declines. Composting accelerates during this stage, and if left unchecked the pile can reach nearly 170°F. Once most of the waste has been digested by the thermophilic microbes, they begin to die off. The temperature then decreases, mesophilic microbes take over again and the compost cycle is essentially complete.

The insulation provided by the layers of compost retains the heat generated by the microbes. The interior of the pile is hottest. While high temperatures aren't essential for composting, they encourage faster breakdown and sterilize the compost. For example, maintaining a pile at 130°F for three days should be enough to kill most of the insect larvae, weed seeds and pathogens that may be present in the interior of the pile. You can destroy those that survive in cooler portions by turning and mixing materials at the edge into the center of a new pile.

Carbon-to-nitrogen ratio—The ratio of carbon to nitrogen (C:N) in organic wastes is especially important for sustaining the microbes. (The microbes use carbon in respiration and to build body tissues, and nitrogen for protein synthesis.) A C:N ratio of about 30:1 provides a good nutrient mix for efficient composting. Lower C:N ratios result in loss of nitrogen to the air as ammonia, while higher ratios result in longer composting times. You can usually attain this optimal C:N ratio by combining several types of wastes. For example, leaves with a C:N ratio of roughly 60:1 can be composted alone, but adding materials higher in nitrogen, such as fresh grass clippings, manure, cottonseed meal or blood meal, speeds the leaves' decomposition.

Particle size—The smaller the particles, the faster organic wastes decompose because there's more surface area exposed to the microorganisms. In particular, smaller particles of woody wastes such as branches and bark break down more quickly. Leaves and coarse weeds also benefit from shredding. Compost made from shredded material has a finer texture and is easy to incorporate into soil.

Aeration—Aerating compost piles is essential for replenishing oxygen consumed by microbes and for venting excess heat and water vapor. Compacted piles compost slowly because of lack of air space. Aerobic conditions can be maintained by turning the pile.

In the absence of oxygen, organic materials can be decomposed by microbes that don't require oxygen (anaerobic microbes), a process that produces compost that's excessively acid. Anaerobic compost piles commonly develop undesirable odors from compounds such as hydrogen sulfide, which smells like rotten eggs. Under some conditions, methane gas is also generated. In aerobic compost, potentially malodorous compounds are oxidized into compounds that don't smell so bad. In addition, compost formed under aerobic conditions retains more nitrogen, since less of it evaporates as ammonia.

Moisture—Composting works best when moisture levels are between 50% and 60% (about as moist as a squeezed-out sponge). Less than 40% moisture inhibits microbial growth, while more than 65% moisture displaces air from spaces in the compost and produces anaerobic conditions. Water is usually added to a new pile or to one that's being turned. As composting proceeds, water evaporates and the pile dries out.

The finished compost—Once composting is complete, the organic wastes are relatively stable. The pile has been reduced in volume by 20% to 60%, the moisture content to less than 40% and the weight by up to 50%. The pH of finished compost is near neutral. Undesirable odors of the starting material are generally replaced by an earthy smell. ☐

Dr. Reginaldo Saraceno is a chemist in the Department of Soil and Water at the Connecticut Agricultural Experiment Station in New Haven, Connecticut.

Cover Crops
Using plants to build the soil

Cover crops protect soil from weather damage—erosion by wind and rain, baking by the sun. Vetch (foreground), a legume, fixes nitrogen in the author's North Carolina garden. Kale, growing behind the vetch, provides young, tender leaves for eating, before being turned under in spring as green manure.

by Louise Langsner

I grow soil in my garden. Flowers, fruits and vegetables grow there, too, but the most important product of my gardening endeavors is the soil. My garden motto for the last 15 years has been "Keep the soil covered." Cover crops help in the year-round effort to combat conditions found on our southern Appalachian farm (in USDA Zone 6): sloping garden sites, thin topsoil, heavy clay subsoil, and generous rainfall punctuated by frequent "gully-washers."

Weather damages exposed soil. I've seen our clay soil turn to brick in the sun and a nearly level field become a miniature Grand Canyon during a heavy rain. Cover crops can help avert both disasters. A dense layer of living plants shades the soil and prevents excessive drying and surface baking. The leaves break the force of falling rain, protecting the soil from compaction, and the roots hold soil in place against erosion.

In addition to preserving physical properties of the soil, plant cover creates a more desirable environment for soil organisms. Rather than seeking refuge deep underground, earthworms and other soil builders stay closer to the surface where they can benefit plants.

Cover crops not only enhance conditions for soil-building organisms, but also build the soil. Contributing nutrients and organic matter, a cover crop is both fertilizer and soil conditioner. The roots often contribute as much organic matter as the tops. Decomposing roots provide air spaces in the soil, thus improving its texture; and soil texture plays a key role in both nutrient and water retention.

"Catch crop" is an English term for a short-term cover crop, and is a reminder of the crop's ability to recycle essential plant nutrients before they're leached from the soil by rainfall. Cover crops catch and store nutrients until these can be turned back into the soil to be used by the next crop.

How to use cover crops

I use cover crops not only over the winter to protect the soil but also throughout the year as green manure, a crop grown for the organic matter it adds to the soil when turned under. If garden space opens up during the growing season, I plant something in it. This technique is especially useful in a large garden such as mine, where the supply of compost is always short of the demand. (You don't need a large garden to use cover crops. My total garden space is a third of an acre, but much of that is in 4-ft. by 16-ft. raised beds. Even city dwellers usually have that much space.)

In spring I use cover crops to rejuve-

Photos: above, Drew Langsner; all others, Thomas E. Eltzroth

A selection of cover crops

Cover crop	Seeding rate (per 1,000 sq. ft.)	Growing conditions	Planting season	Comments
LEGUMES Austrian winter pea (*Pisum sativum* var. *arvense*)	2-4 lb.	Cool weather; winter-hardy	Early spring or early fall	Good companion to small grains. Makes most growth in spring. Fixes nitrogen (N) at 70-125 lb./acre. Good to excellent yield of organic matter.
Azuki bean (*Vigna angularis*), bush bean (*Phaseolus* spp.)	3-4 lb.	Warm soil; shade-tolerant	Summer	Sow densely for good weed competition. Underplant in corn. Edible beans. Fixes N at 80 lb./acre.
Cowpea (*Vigna unguiculata*)	2-3 lb.	Warm soil; moderately shade-tolerant	Summer	Underplant in corn, orchards or vineyards. Edible beans. Fixes N at 80 lb./acre.
Crimson clover (*Trifolium incarnatum*)	½-2 lb.	Winter-hardy; shade-tolerant	Early spring through early fall (plant by early fall for adequate growth)	Underplant in corn. Fixes N. Very good to excellent yield of organic matter. Beautiful bloom; reseeds easily.
Fava bean (*Vicia Faba*)	2-4 lb.	Cool weather, moist soil; hardy to about 10°F	Early spring or late summer	Strong, deep root system. Edible beans. Fixes N at 150 lb./acre. Excellent yield of organic matter.
Garden pea (*Pisum sativum* vars.)	2-4 lb.	Prefers cool, moist soil; tolerates light frost	Early to late spring, or August	Edible peas. Fixes N.
Soybean (*Glycine Max*)	2-3 lb.	Warm soil; needs good drainage and regular moisture; not frost-tolerant	Summer	Underplant in corn. Garden varieties produce edible beans. Fixes N at 100 lb./acre. Good to excellent yield of organic matter.
Vetch, common or hairy (*Vicia* spp.)	1-2 lb.	Winter-hardy; makes most growth in spring	Late summer to early fall	Interplant with small grains. Fixes N at 125 lb./acre. Honey plant.
GRAINS Barley (*Hordeum vulgare*)	4-6 lb.	Winter-hardy	Early to mid-fall	Good companion to winter-hardy legumes. Excellent yield of organic matter. Straw crop.
Buckwheat (*Fagopyrum esculentum*)	2-3 lb.	Warm soil; will not tolerate frost	Summer; 2-3 crops possible per season	Excellent weed competition. Will grow on almost any soil. Excellent yield of organic matter. Reseeds very readily. Honey plant.
Oats (*Avena sativa*)	3-4 lb.	Less hardy than rye or barley	Early fall for winter cover; early spring for green manure	Good to excellent yield of organic matter. Good straw crop.
Ryegrass (*Lolium* spp.)	1 lb.	Cold-tolerant but not winter-hardy	Early spring to late summer	Root system provides good erosion control and soil conditioning. Turn under at 6 in. or mow to keep short.
Winter rye (*Secale cereale*)	4-6 lb.	Winter-hardy	Fall	Vigorous grower. Best crop for winter soil protection. Excellent yield of organic matter. Straw crop.

nate the raised beds that have held over-wintered salad greens. By March or April I chop in the remains of the crop with a dressing of manure or compost. I then sow leftover seed of cool-weather plants such as mustard, kale, radishes, peas, clover, fava beans or oats to grow until the bed is needed later in the season. One benefit of these green-manure crops is that they're easily turned under. Similarly, as spring crops are harvested, I plant a quick-growing, warm-weather catch crop such as buckwheat, cow peas, azuki beans or soybeans until the ground is needed for fall vegetables.

Corn is well suited to companion cover crops. I plant corn in double rows 12 in. apart. Each set of double rows is spaced 42 in. apart—the width of my tiller. When the corn is knee-high, I broadcast crimson clover throughout the patch. You can also use cow peas or azuki beans this way. These covers crowd out weeds and provide green manure. Crim-son clover is hardy and remains in the garden to serve as a winter cover after the corn has been harvested.

Selection and planting

In addition to clover, vetches and winter peas are hardy cover crops that may be companion-planted into the fall garden. I usually broadcast the seed into beds of late-season crops or summer vegetables nearing the end of their production in early September. This works best around upright plants such as tomatoes, peppers, beans, trellised cucumbers or eggplant. With a hoe, I chop the soil lightly around the plants to cover the seed. If there isn't enough moisture in the ground for the seed to germinate, I use a sprinkler to get the crop growing. Interplanting these slow-growing legumes with the end-of-season garden vegetables, before the garden is ended by frost, is important to allow the cover crops to make adequate growth before cold weather sets in.

I prefer sowing my cover crops in September, but any time through October is still all right. It's generally the end of September before we get around to digging potatoes and sweet potatoes and gathering winter squash and pumpkins. Then I clear the garden of any plant debris that might host insect pests or disease, and till in mulch, weeds and plant roots to make a level seedbed. Finally, I broadcast seed over the area and rake to cover it. Fall rains do the watering, and the seedlings ordinarily emerge within a week.

I interplant a small grain (my favorite is barley, but I also use winter rye and oats) with winter peas or vetch. The grain makes good fall growth; then the legume takes off in spring. Legumes fix nitrogen in the soil, but are more expensive than, and not as hardy as, grains. Combining the two assures you of a crop even if the winter is very cold. A good rule of thumb

Mustard will grow anywhere, and its large root system loosens even compacted ground. In addition, its brilliant yellow flowers are a delight to the eye, and the young, tender leaves are edible.

Choices for spring planting

Peas—These plants germinate in very cool soil, so I use them on beds that have overwintered greens or carrots and are all but empty by the end of February. Austrian winter peas or 'Alaska' (pea soup) peas are the most reliable for very early planting, as they withstand late freezes and spring snowstorms without damage, but you can use other garden varieties. I broadcast the seed and rake it in. Peas prefer a soil pH above 6.0 and a good supply of phosphorus and potash. They can fix 70 lb. to 125 lb. of nitrogen per acre. Cut back the succulent vines and add them to the compost pile; or leave them on the plot, where they quickly decompose when turned into the soil. Added benefits of a pea cover crop are the peas, which can be harvested (except for Austrian peas), and the tender leaf tips and flowers, which are delicious in spring salads.

Fava beans—Sometimes called broad or bell beans, fava beans are cold-tolerant and can be sown in very early spring. I grow them in south-facing garden beds that warm up quickly. Plant seeds 1 in. to 2 in. deep, spaced 6 in. apart each way. Favas have strong, deep roots that bring up nutrients and open the soil. In addition, they can fix up to 150 lb. of nitrogen per acre. My favas grow to 3 ft., but I've seen fall-planted favas in California that grew as tall as a man. Fava beans can be eaten fresh, or can be dried; also, the flowers are attractive to bees. Mow or scythe the plants before working them into the soil. Favas are not always winter-hardy in my region, but I sometimes plant them in late summer to make a fall crop, then leave the frozen plants in the garden as a winter mulch.

Annual ryegrass—Annual ryegrass is a quick-growing, vigorous cover crop that germinates in cool soil and is very cold-tolerant. It has a large, fibrous root system that is beneficial both in loosening heavy soil and in protecting against erosion. Sow ryegrass from early spring to late summer; rake the seed to cover it. Ryegrass winter-kills here, but the roots and wilted tops still hold and protect the soil.

Oats—Planted in early spring as a fast-growing green-manure crop, or in early fall with legumes as a winter cover, oats are less vigorous and cold-hardy than winter rye. I grow them mainly when I want straw or seed heads for making wreaths. A mature oat crop can add a lot of organic matter to the soil; one memorable crop of oats that we allowed to head out before disking it into the soil broke down to make a rich, dark seedbed that looked as though we had spread compost over the whole field.

Mustard—Many varieties of mustard are useful as cover crops because they grow quickly and can be used from mid-spring

is to sow half the recommended rate of each for the particular size plot.

Turning cover crops under

Cover crops are welcome patches of green throughout the winter, and really brighten up and take off in growth as spring arrives. I let them grow as long as possible before turning them under. I begin turning some of the crops under in March, double-digging the raised beds and tilling the conventional gardens. I leave other areas of the garden until April or May.

It's best to turn under barley, oats or rye when the crop is about 6 in. high, while the clumps are still tender and easily broken up. If you can't turn in the crop at 6 in., mow it back. Taller, older growth is unmanageable, tough, and hard to incorporate into the soil; it may even keep growing after being turned under. Legumes are not so apt to get out of control, as their growth is not so rampant. Still,

scythe or mow back legumes when they get to be 1 ft. or so long. Turn in buckwheat as it begins to flower, or you'll have it everywhere.

Turned-under cover crops need time to break down before anything else can be planted. How long depends on age and type of plant, weather, soil temperature, and amount of biological activity in the soil. The time needed ranges from a week to a month. To hasten the process, I sometimes harvest the plant tops for the compost pile and work only the roots into the soil.

There are many things to consider when choosing a cover crop for your garden. Most important are the time of year a cover is wanted and how long a given piece of the garden can be left fallow. Other factors to weigh include growth rate, potential production of organic matter, root depth and structure, benefit to bees, the crop's mineral content, and its nitrogen-fixing capacity.

until late summer. Mustards will grow anywhere—including pathways and other compacted ground—and their large root systems loosen heavy soil. Agricultural varieties of mustard probably provide more bulk and have deeper roots, but I often use garden varieties because I usually have a supply of the seeds. In addition, garden mustard is milder in flavor, and we enjoy the young leaves in salads or stir-fries, then let the plants mature to flowering stage before turning them under. I broadcast the seed and rake lightly to cover it. Seedlings get thinned (and eaten) to 4 in. to 8 in. apart, depending on the variety and how long the plants will be left to grow.

Choices for summer planting

Buckwheat—A tender annual, and one of the best choices for a warm-weather cover crop, buckwheat should be planted well after the last frost. Broadcast the seed and rake it in. It germinates easily and grows quickly; the crop is ready to be turned under after only 30 to 40 days, when it flowers. Quick growth makes buckwheat excellent for smothering out weeds, and allows for a good green-manure crop when a plot is available for only a few weeks. Buckwheat produces lots of organic matter that breaks down easily, and it's an outstanding bee crop as well. It self-seeds readily and can become a weed (though not a pernicious one).

Soybeans—Planted after the soil is warm and all danger of frost is well past, soybeans will thrive. They do best in fertile soil, though they grow on almost any type of ground; interplant them with buckwheat if the soil is poor. Soybeans can fix up to 100 lb. of nitrogen per acre and reach the flowering stage (the time to turn them under for maximum nitrogen fixation) in about 60 days. Any variety will make a good cover crop, but agricultural forage soybeans provide the most bulk. Soybeans are an excellent source of organic matter for the garden, but the stems and root system are tough; use a tiller when working them into the soil.

Other legumes—Cowpeas, azuki beans and other annual legumes are adapted to a wide range of soils, and can fix up to 80 lb. of nitrogen per acre. Planted thickly, they'll smother out most weeds. Beans are also well suited for undersowing in corn, being moderately shade-tolerant. When turned in young, these plants break down fairly quickly; a thick stand makes a good mulch if allowed to mature.

Winter-hardy cover crops

Crimson clover—Like all clovers, this type is a good nitrogen fixer with a strong root system that provides excellent erosion control. Crimson clover germinates in warm weather; broadcast seed and rake it in anytime between spring and early

Fava beans really open up difficult soil. The strong, deep root systems penetrate heavy clay, breaking it up and allowing air and water to enter. The flowers attract bees, and the edible beans are a bonus.

Buckwheat is a good warm-weather crop for smothering weeds. Its hollow stems and succulent, tender leaves break down quickly. Turn it under before seed sets to prevent it from self-sowing.

SOURCES

The following mail-order companies sell seeds of a wide variety of cover crops:

Bountiful Gardens—Ecology Action, 5798 Ridgewood Rd., Willits, CA 95490. Catalog free.

Johnny's Selected Seeds, Foss Hill Rd., Albion, ME 04910. Catalog free.

Necessary Trading Company, New Castle, VA 24127. Catalog $2.00.

Peaceful Valley Farm Supply, P.O. Box 2209, Grass Valley, CA 95945. Catalog $2.00.

fall. This clover is also shade-tolerant, and can be sown under corn or other upright vegetables, in the orchard, or among berry bushes. It's winter-hardy, but I plant it by early September here (about a month before frost) to get an adequate cover before cold weather. In May and June the bright-red blooms make quite a show. Plants reseed easily.

Austrian winter peas—See the description under spring choices. Plant in early fall.

Vetches—These viney legumes fix up to 125 lb. of nitrogen per acre, smother out weeds, and provide large amounts of organic matter. Plant vetch in early fall, preferably with rye or barley to make a better winter cover. Hairy vetch is recommended as the most winter-hardy in my area, but I use common vetch with good results. Late-spring growth can get tough; mow or scythe the vines to keep them under 18 in. long. Vetch's beautiful purple flowers attract bees.

Winter rye—A small grain that will make a stand on just about any type of soil, winter rye is the most easily established, reliable winter cover crop I know of. It's vigorous, quick-growing and very hardy. Perhaps a little too vigorous. It may crowd out interplanted legumes, or grow so fast in the spring that it gets ahead of you—mow it back if necessary. The extensive, fibrous root system prevents erosion, and adds a lot of organic matter to the soil. I plant rye from early September to the end of October. If allowed to mature, rye makes excellent straw for mulch.

Barley—I prefer barley to winter rye because it has the good qualities of rye without the drawbacks. Barley grows a bit more slowly, so I can't plant it as late, but it's less likely to overtake interplanted legumes. It's easier to control in the spring as well. □

Louise Langsner gardens in Marshall, North Carolina.

CLOVER AS GROUND COVER AND COMPOST

by Jonathan Frei

I garden for a fitness spa in Baja California, Mexico, where my job is to grow organically as many of the vegetables, herbs and fruits used in the spa kitchen as possible. We use wide, raised beds that are separated by 19-in.-wide paths, just ample enough to accommodate my mower and wheelbarrow.

These pathways comprise 40% of my total garden area. Unfortunately, weeds thrive in bare paths, and I used to devote a lot of time to maintenance. It's one thing to have bare pathways, and another to keep them that way. Also, I didn't like leaving bare soil continually exposed to the elements. Because our soil has lots of sand but only 1% or less organic matter, it can be too hot to kneel on at midday in summer. I solved these problems by sowing Dutch and New Zealand white clovers, both of which are varieties of *Trifolium repens,* in the paths. Both are low-growing perennials that tolerate regular, close mowing. Of the two, Dutch white clover grows lower, while the New Zealand tolerates more heat.

The clover-covered paths provide me with many benefits, though I still spend time maintaining them. Instead of weeding, I mow the paths, collect the clippings in a grass catcher and add them to the compost. The finely chopped, nitrogen-rich clover is excellent for heating up a compost pile. The clover covers the paths year-round, and I never have to reseed it or cultivate the soil. And, like other legumes, clover fixes nitrogen, as much as 160 lb. per acre. To take full advantage of this trait, I inoculate seed with *Rhizobia* bacteria, or purchase pre-inoculated seed. The bacteria are necessary for the legume to convert atmospheric nitrogen to a form usable by plants. They live in small nodules attached to the roots of the legume.

Our climate is influenced by the proximity of both desert and coast; hot days and cool nights are the norm (USDA Zone 8 conditions). White clover grows actively here about 40 weeks of the year, slowing down from early December to late February. From March through November, it grows 1½ in. a week. I've calculated that, over a year's time, I've cover-cropped more than an acre of my garden to a height equivalent to 5 ft. By composting this nitrogen-rich resource and applying it to my vegetable beds, I can keep the beds in production longer before I have to plant a cover crop to enrich them. This compost also reduces my need for outside nitrogen sources (typically, dairy manure), many of which have high sodium levels.

Water is too scarce here to grow clover in all the paths—our annual rainfall is normally 15 in. from fall through spring—so we concentrate on a one-acre area where beds

Clover paths protect soil from erosion, crowd out weeds and provide compost material. The author edges the paths three or four times a year with a spade, and adds the clover clumps to his compost pile.

Paths are 19 in. wide—just the width of the mower. A grass catcher collects clippings for compost.

of salad greens and other cool-weather crops are in continuous production. The beds are watered by drip irrigation, and we use overhead sprinkling every seven to ten days during the summer to keep them green. Last year we received only 8 in. of rain, and had to discontinue overhead watering, but the clover on the edges next to drip-watered beds continued to grow, and in the fall it grew back out into the center of the paths and soon rejuvenated itself.

Clover readily spreads and roots into the beds. There are advantages to this invasion. Raised beds are more susceptible to drying out than conventional beds because the edges are exposed to wind and sun. When clover grows up the edges of our beds, it functions as a living mulch. Normally the edges of a raised bed dry out first, but with clover growing up the sides, they stay moister longer than the centers. I can usually leave the clover for four to five weeks before it begins to send down roots.

To contain the clover, I just grab fistfuls and pull it out of the beds. Or, I use a shovel to edge the clover when I replace the annual crops in the beds (three or four times a year). Over time, that adds up to a significant amount of clover added to the compost pile.

The clover also helps tremendously in reducing soil temperature and raising humidity around the seedbeds. This is especially welcome in August, when I have to sow fall and winter crops of beets, carrots and cabbage in the heat.

Having so much of the garden permanently covered helps protect the soil from heavy winter rains, minimizing compaction and nutrient leaching, and eliminating muddy conditions. It also reduces compaction from foot and wheelbarrow traffic. Clover increases the earthworm population, but it can harbor snails and slugs as well. In wet climates where these pests are a problem, try close cutting and regular edging. I've frequently cut the clover as close as ½ in., and it's always returned with vigor.

With the first fall rain, I sow annual subclovers (*Trifolium subterraneum*) in other parts of the garden. These flower very close to, if not right on, the ground, and they reseed prolifically. I can mow as close as I wish, and they still reseed. The soil is protected during our winter rains, and because subclovers are better cool-weather growers than the perennial white clovers, they provide five months of cutting for compost. In spring the subclovers reseed and die, leaving a protective layer over my soil. They germinate once again with the fall rains. ☐

Jonathan Frei is head gardener at Rancho La Puerta, in Baja California, Mexico.

Photos: Robert Kourik

Microscopic Partnership
Fungi that help plants grow better

by Suzanne M. Schwab

When gardeners think about fungi that grow on plants, they usually think of problems such as mildewed roses, blighted potatoes and curled peach leaves. But not all fungi are harmful. In fact, plants commonly benefit when fungi colonize their roots and form root/fungal associations called mycorrhizae. Particularly in infertile soil, plants with mycorrhizae may photosynthesize more rapidly, tolerate salinity better, resist competing root pathogens and take up water more efficiently than uninfected plants do.

There are several distinct kinds of mycorrhizae. I study vesicular-arbuscular mycorrhizae (VAM), which occur on as many as 90% of all flowering plants. My research explores the effects of inoculating plants with VAM fungi. Inoculation promises many advantages, and at least one company offers an inoculant product. But inoculation isn't a panacea. Sometimes it brings benefits, and sometimes it doesn't. I'll tell you about what biologists have found out so far, then, since VAM is an ideal subject for backyard research, I'll outline a method you can follow to inoculate plants in your own garden and observe the results.

What is it?
The VAM association is a partnership. The fungus, which cannot manufacture its own food, penetrates into the cells of the host plant's roots, where it absorbs sugars and other organic nutrients. At the same time, it absorbs essential minerals, especially phosphorus, from the surrounding soil, and transports them back to the roots. There, the minerals that the plant needs are somehow "traded" for the carbohydrates that the fungus needs.

VAM fungi occur naturally in the upper few inches of soil wherever plants are growing, but they can't exist without plants. The fungi usually are absent from soil that's been recently flooded, and from sites where the topsoil has eroded or been removed. Several closely related species of fungi can form VAM associations, but they differ in their ability to take up minerals. Unlike some kinds of fungi that associate with plants, VAM fungi are remarkably promiscuous—any given species may colonize the roots of thousands of different kinds of plants. In fact, a single VAM fungus may bridge the roots of two plants of different species if they're growing next to each other.

VAM are invisible to the naked eye, but with a microscope you can see the fungal spores and hyphae that form outside the plant's roots (photos p. 46). Hyphae are thin filaments, usually less than $1/250$ in. in diameter, that constitute the basic structural and functional units of a fungus.

The hyphae of a VAM fungus penetrate into the cells of the cortex, or outer layer, of rapidly growing young roots. There they form two characteristic structures: vesicles and arbuscules. Vesicles

> **In effect, mycorrhizae greatly increase the plant's root system.**

develop when the tips of hyphae swell into small round sacks that store fats. They're formed most abundantly at the end of the growing season. Their high fat content has led most researchers to assume that vesicles are a food-storage structure for the fungus; but, frankly, nobody knows how long they last or what they're really doing.

Arbuscules form when hyphae penetrate into a root cell and begin to branch repeatedly. A hypha can penetrate a root-cell wall and form an arbuscule in just a day or so, but arbuscules begin to deteriorate within three to four days and normally disintegrate within a week of their initial formation. It's not unusual for a single root cell to contain more than one arbuscule in different stages of development. Arbuscules are involved in the transfer of phosphorus (and other minerals) to the plant, and carbohydrates to the fungus, but nobody knows just how this transfer takes place.

Outside the root, the tiny threadlike hyphae of VAM fungi may reach as far as $2\frac{1}{2}$ in. into the soil. In effect, they greatly increase the extent and surface area of the plant's root system. This is especially important for the uptake of phosphorus, and in most cases the enhanced growth of plants with VAM is due to the increased uptake of phosphorus. Phosphorus moves through the soil very slowly, and a major limitation to its uptake is the rate at which it diffuses to the root surface. VAM fungal hyphae can absorb phosphorus and transfer it (via cytoplasmic streaming) to a plant's roots much more quickly than phosphorus moves through the soil on its own. Moreover, the fungi appear to take up phosphorus in the same form preferred by plants, i.e., as a phosphate ion.

What we know—and don't know
My current research illustrates a few of the many puzzles to be solved in predicting the effect of mycorrhizae on plants. For the past four years, I've been working with the Washington State Federated Garden Clubs on their "Operation Wildflower." This project involves members throughout the state who collect native wildflower seeds, then raise seedlings and transplant them at highway rest stops and sites in need of revegetation following road construction. I'm testing several species of wildflowers to see if their survival and growth after transplanting are improved if the seedlings are inoculated with VAM fungi. So far I've tested 20 species of plants. Six species have shown improved growth in the greenhouse if they were inoculated with my isolate of a VAM fungus. Four of these maintained this advantage after transplanting, and one appears to be absolutely dependent on VAM to survive in the field.

These results raise many questions. Would the 14 species that don't benefit from my VAM fungus do better with another species of fungus, or do they not respond to VAM fungi at all? What about the two species that benefit in the greenhouse but not in the field—do they lose their VAM fungus when they're transplanted? Or do non-inoculated plants become colonized by an indigenous fungus that allows them to catch up to their inoculated counterparts? Does the fungus spread from inoculated to non-inoculated plants once they're transplanted into the same soil, and, if so, how quickly? And do the plants that do benefit from inoculation eventually become colonized by native VAM fungi after transplanting? If so, how do the inoculated and the native fungi compete?

A second project of mine, testing the response of a native prairie grass and a weedy "cheat" grass to VAM, reveals still more complications. When grown separately, the weedy grass usually is stunted when I add VAM fungi to the soil, while the native grass grows more rapidly when VAM fungi are present. If I grow the two grasses together in the same pot, however, the relationship reverses. Then the weed is benefited by the VAM fungus and the native grass is harmed. Apparently, plants respond differently to VAM if grown in monoculture or in competition with other plants.

Other researchers offer related observations. Within a single plant species, there can be considerable variation in how different cultivars respond to VAM. In one experiment, one variety of citrus grew more than twice as fast when inoculated with a VAM fungus, while a second variety hardly responded at all.

It's hard to sort out all the environmental interactions that can affect the efficiency of a given strain of VAM fungus and modify how a given plant will respond to the fungus. Because the primary effect of VAM is to enhance mineral uptake, it's not surprising that plants generally are less likely to benefit significantly in soil that is already quite fertile. After all, the plant does have to supply the fungus with organic nutrients at its own expense. In fertile soil, the plant may be able to obtain its necessary mineral nutrients just as well without the VAM and save itself the carbohydrate cost of supporting the fungus. In fact, plants seem to somehow protect themselves against being colonized by the fungus in very fertile soils.

Besides soil fertility, VAM effects can be influenced by light levels and day length, temperature, air pollution, soil moisture levels, soil pH and salinity, the application of pesticides, and the presence of plant pathogens. Most of these factors affect both partners in the mycorrhizal association—the plant and the fungus. Generally, VAM are most likely to be beneficial to a plant stressed by suboptimal conditions below ground, such as low soil fertility or moisture. In these cases, the fungus acts as a supplement to the plant's root system. VAM are less helpful to a plant stressed by above-ground factors, such as low light or insect predation. In these cases, the fungus is a drain on the plant's already limited carbohydrate supply.

VAM are also influenced by other soil microorganisms. This may be why plants

MYCORRHIZAE UP CLOSE
Magnification: 14X
Hyphae
Root
Fungal spores

Although invisible to the naked eye, vesicular-arbuscular mycorrhizae (VAM) show up clearly under the microscope. The photos here show VAM fungi growing in association with the roots of a leek plant. At low magnification, the leek roots look like translucent gray ribbons (above). The slender hairlike filaments that surround the roots are the hyphae of the VAM fungus. These absorb phosphorus and other nutrients from the soil. The dark spherical objects are the fungal spores.

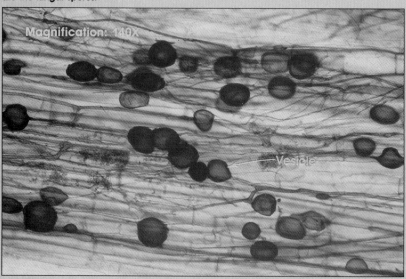

Magnification: 140X
Vesicle

When the VAM fungus penetrates inside root cells, it forms characteristic structures, visible in these photos of specially prepared leek roots. (The leek roots have been "cleared" so that the cells look like glass bricks.) Vesicles (above) are the dark round objects inside the cells. They store fats and organic nutrients absorbed by the fungus from the plant. Arbuscules (below) are brushlike structures that form when fungal hyphae penetrate a root cell and branch repeatedly. They're involved in the transfer of materials between the fungus and the plant.

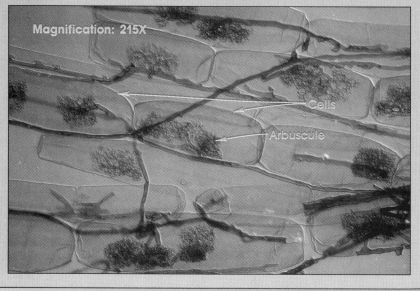

Magnification: 215X
Cells
Arbuscule

All photos, except where noted: Mark Brundrett

Both of these blanketflower seedlings have grown for three months in a greenhouse. The plant on the left was inoculated with author Schwab's pot culture of VAM fungus; the one on the right wasn't.

that show a spectacular response to VAM in the greenhouse or in fumigated soils may be a disappointment in non-fumigated soils. Soil-inhabiting insects may "graze" on the mycorrhizal hyphae that extend out into the soil and do all the mineral-absorbing, so that the beneficial effects of the fungus are lost. Research in Kansas suggests that certain bacteria also suppress VAM by an unknown mechanism. Finally, different VAM fungi may compete with each other. Unless the fungus used for inoculation is more efficient than the indigenous fungi, inoculation will have no beneficial effect. In fact, if the inoculated fungus is less efficient, it could be detrimental by competing with the more efficient native fungi on the root. On the other hand, even a more efficient inoculated strain of fungus may not survive in the field as well as less efficient but hardier native strains. As you see, mycorrhizal research can be perplexing.

Gardening with VAM

Would adding VAM help the plants in your garden? Under certain conditions, inoculation with VAM fungi benefits many cultivated plants—including apple and grape seedlings, citrus, sweet gum and ash trees, many legumes, pasture grasses, onions, lettuce, corn, wheat, barley, strawberries, peppers, and even tomatoes. VAM don't form on members of the cabbage family, however, or on spinach, beets, buckwheat and amaranth, or their relatives. I'd suggest that you're most likely to see a difference between mycorrhizal and non-mycorrhizal plants with slow-growing perennials, especially in infertile or eroded or disturbed soil. If you're considering a new landscaping project on bare soil, you might try adding VAM fungi as an alternative or a supplement

to fertilization and irrigation. Remember that the advantages of VAM inoculation diminish in a well-manured garden, because VAM fungi may already be present there, and because the benefits due to VAM diminish as soil fertility increases.

To set up an experiment and explore the effects of VAM, all you need to do is acquire a supply of VAM fungus, apply it to your plants, and compare their growth to other plants that are grown at the same time and in the same conditions

> **You might try adding mycorrhizae as an alternative to fertilization.**

without VAM. It's fairly easy to grow your own isolate of VAM fungi. Most soils that have not been disturbed recently have some mycorrhizal fungi in them. An unfertilized grassy meadow or pasture or a relatively dry shrubby area would be likely harvesting sites. Collect a couple of scoops of soil from under the roots of dominant plants, along with whatever root fragments are present in the soil. You won't see any sign of the fungus without a microscope, but if you've selected a good soil it's there.

To increase the amount of fungus, you can propagate it in what is referred to as a pot culture. Fill one or two clay pots with a mix of about 7 parts sand to 1 part potting mix, sandwiching about a 1-in.-thick layer of your collected soil in the middle of the pot. Since VAM fungi grow on most plants, you have a wide choice of hosts to use to grow your fungus, but it's best to use a host plant for your pot culture that is not related to whatever you

hope to inoculate. That way, you're much less likely to be growing harmful organisms at the same time you're growing the VAM fungus. I like to use coleus as a pot-culture host plant because as a member of the mint family it's not closely related to anything else (except basil) that I'll be inoculating. Whatever you use as the host, remember that your goal is to grow lots of VAM fungus, not necessarily a beautiful pot-culture plant. The fungus will grow and produce spores more prolifically if you give the plant optimum lighting but keep the soil infertile and a bit drier than the plant would prefer. It also helps to trim off any flowers that might form on your host plant so that seed production doesn't compete with the fungus for the plant's carbohydrate reserves. In good light, it takes about six months to produce a pot full of mycorrhizal inoculum. After that time, you can store the VAM inoculum, if you wish. Cut off the above-ground parts of the plant, spread the pot-culture soil on a sheet of paper to dry, then store it in a paper sack in the refrigerator.

The "do-it-yourself" method has two drawbacks. First, unless you have access to a microscope and some training in what to look for, there's no way to guarantee that the fungus is actually present in your pot culture; it may never have been present in your original field sample, or it may have died at some point in the pot culture or the storage process. Second, even if the fungus is present, you have no idea how efficient it is in promoting plant growth until you test it.

Using a commercial inoculum avoids the possible drawbacks of producing your own pot culture. Although I haven't tried it yet, I know of an inoculum called Nutri-link that's grown under special conditions to keep it free from fungal pathogens, and that has been tested for efficiency in promoting mineral uptake on a variety of plants under a variety of soil conditions. (Nutri-link is available from Smith & Hawken, 25 Corte Madera, Mill Valley, CA 94941; the cost is $16.45 postpaid for 16 oz.)

Once you have a source of VAM fungus, it's simple to add it to your garden. Add a ¼-in. layer of the pot-culture soil to the bottom of a 1-in.-deep furrow before planting a row of seeds. Or, add about ½ in. of pot-culture soil to the bottom of a transplanting hole before setting a plant in place. You can incorporate a few spoonfuls of pot-culture soil with the potting medium you use to raise seedlings in containers. In any case, remember to grow similar plants in untreated soil as a control. Finally, observe and compare the growth of the treated and untreated plants. ☐

Suzanne Schwab is an assistant professor of biology at Eastern Washington University in Cheney, Washington.

Here are the main soil amendments (from left to right): hydrated lime and ground limestone, which raise pH; sulfur, ferrous sulfate and aluminum sulfate, which lower pH; and peat moss, representing organic matter, which in almost any form—shredded leaves, sawdust, straw, compost—improves soil texture and fertility.

Soil Amendments

How lime, sulfur and organic matter improve poor soil

by Mark Kane

Soil amendments correct two common soil deficiencies. The first is unsuitable pH—too acidic or too alkaline for the plants you want to grow.

The second is poor structure — either too heavy and ill-drained, or too light and quick to dry. To get a pH that makes nutrients most available to plants, you raise it with lime or lower it with sulfur. To improve soil structure, the best measure is applying organic matter regularly. Organic matter transforms clay soil, making it more porous and hospitable to root growth. It also

helps sandy soil hold enough water to keep plants growing during dry spells.

Effects of pH on nutrients

Soil pH is a measure of the acidity or alkalinity of the water in soil. The pH scale runs from 0, which is very acidic, to 14, which is very alkaline. The center of the scale, pH 7, is neutral, and pH weakens as it moves toward neutral from either side. For example, pH 6 is less acidic than pH 5, and pH 8 is less alkaline than pH 9 (see the chart at right). Most U.S. soils have a pH between 5 and 9. In areas of abundant rainfall, including most of the eastern U.S., soils tend to be acidic, since water moving through the soil leaches out alkaline compounds. In arid regions, including much of the western U.S., soils tend to be alkaline.

Soil pH affects the availability of plant nutrients. Phosphorus—a nutrient plants require in considerable quantity—is most available between pH 6 and 7. As the pH drops below 6, or rises above 7, phosphorus combines with other minerals, becoming less available for plants. Below pH 5, phosphorus is scarcely available at all, a condition few plants can tolerate. In acid soil, some nutrients become *too* available. Iron, zinc, copper, manganese and aluminum can become so relatively abundant below pH 5 as to be toxic to most plants.

How do you learn the pH preference of plants? For the most part, reference books and nursery catalogs refer to pH vaguely. They say a mountain laurel prefers "acid soil," or a witch hazel grows in "any good garden soil." You should interpret "acid" as roughly pH 5.5, and "good garden soil" as pH 6 to pH 7. If you can't find a pH preference for the plant you have in mind, you won't go far wrong if you assume it's between pH 6 and pH 7, where nutrients are most available. Most garden plants grow well anywhere in that range. A few woody plants, however, thrive only on acid soils. Rhododendrons, azaleas, blueberrries, cranberries, heaths and heathers, among others, grow best at a pH around 5.5. Indeed, rhododendrons develop iron deficiencies and produce pale or yellow leaves if the pH is much above 5.5.

There are two ways to measure the pH of your soil: you can send a soil sample to the laboratory run by your state extension service, or you can test the sample yourself with a kit, a pH meter, or litmus paper. (See "Home Soil Testing," pp. 16-20, for more information.) Look for the extension service in the government listings of your phone book. The lab fee is usually small, and the test results typically include not only your soil's pH, but also its fertilizer needs, its organic matter content, and its capacity to hold nutrients. Of all the methods for testing pH, using litmus paper is the simplest and least expensive. You moisten a soil sample with water, press the paper against the sample, wait for the paper to change color, and then read the pH from a color chart.

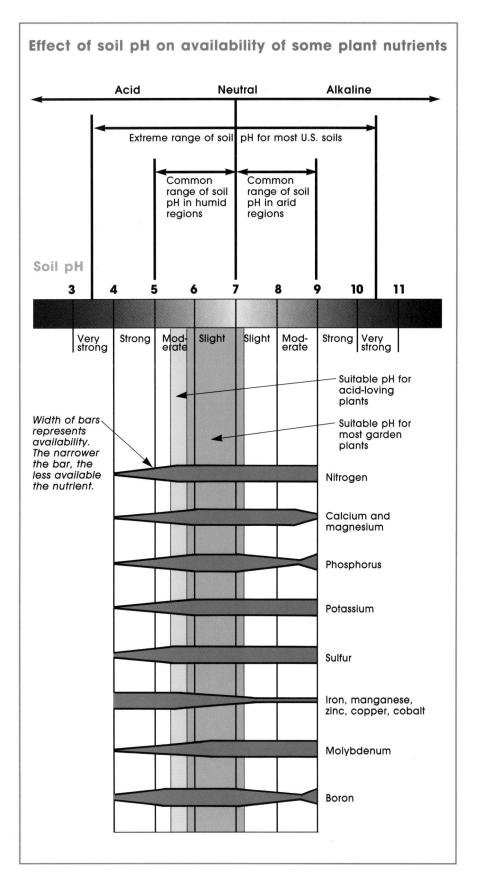

Effect of soil pH on availability of some plant nutrients

If the pH of your soil is unsuited to the plants you want to grow, you have to alter the pH with a soil amendment. You can amend the soil whenever you like, but follow the recommendation of a soil test about the amount of amendment to use. If you have to make a big change in pH—more than one point—the recommendation should tell you to apply half the amendment one year and half the next. If you add the entire amount at once, the pH can temporarily change too much in the top layer of soil.

Adjusting pH with soil amendments

Raising pH—To raise pH, you spread lime and mix it into the soil. Lime for gardening comes in two forms. One is finely-ground limestone. The other is hydrated lime—limestone that has been crushed, heated and milled. Gardeners tend to call both "lime," though ground limestone is slightly less potent than hydrated lime and alters pH more slowly. In some regions, gardeners lime with ground oyster shells or ground marl, a soft form of limestone. (Wood ashes also raise pH, but they are most useful for their nutrients.) Ground limestone, which has the consistency of sugar, is inexpensive and easy to use. It is sold in 50-lb. sacks by garden centers and in bulk by farm centers. Hydrated lime, which has the consistency of talcum powder, is sold in 40-lb. sacks and smaller quantities by most garden centers, and also by masonry supply shops (where it may be called slaked lime or builder's lime). Hydrated lime is more alkaline than ground limestone, so smaller amounts are needed for the same change in pH. Use it at three-quarters the rate of ground limestone. Whichever lime you use, spread and mix it evenly, avoiding local concentrations that might harm roots near the surface.

In some regions of the eastern U.S., where soils are acid and deficient in magnesium, gardeners use dolomitic lime—ground or hydrated—to raise the pH and to amend the magnesium deficiency. Dolomitic lime means "more than 35% calcium magnesium carbonate." Your local extension service will be able to tell you whether it suits your conditions.

The amount of lime you need to apply depends on soil and climate. Clay soils need more limestone than sandy soils, and limestone works faster in warm, wet climates than in cool, dry climates. In the north, raising the pH one point takes 3 lbs. to 12 lbs. of limestone per 100 sq. ft.—3 lbs. for sandy soils, 12 lbs. for clay soils. In the south, the amounts are slightly less.

Besides raising pH, liming also has an effect on the texture of the soil. It binds clay particles together in crumb-like groups, making the soil slightly more granular and porous. It also makes sandy soil pack tighter, slowing drainage.

Lowering pH—To lower pH, mix acidic minerals into the soil. The most commonly used minerals are sulfur, ferrous sulfate and aluminum sulfate. Sulfur, milled to a powder-like consistency, is sold in small quantities by most garden centers, and in bulk by many farm centers. Sulfur is also an effective fungicide when dusted on leaves, and some garden centers sell sulfur only in small quantities as a fungicide. Sulfur is, pound for pound, considerably more effective at lowering pH than ferrous sulfate and aluminum sulphate.

The other acidic minerals are less potent than sulfur, but more often found at garden centers in quantity. Ferrous sulfate (also called iron sulfate and copperas) adds iron to the soil. If an acid-loving plant has yellow foliage

The pale green of these rhododendron leaves is a sign of iron deficiency, most often caused by an insufficiently-acid pH. Adding sulfur to the soil will lower the pH, make iron available, and restore the leaves to their characteristic deep green color.

from lack of iron, ferrous sulfate can sometimes produce deep green foliage quickly. Aluminum sulphate reduces pH, but also adds aluminum to the soil in considerable quantities. If you choose someday to remove your acid-loving plants and replace them with plants that prefer a higher pH, they may suffer from aluminum toxicity. (Ferrous sulfate and aluminum sulfate are available from Mellinger's: 2310 W. South Range Rd., North Lima, OH 44452-9731.)

To lower pH, spread 1 lb. of sulfur, or 5 lbs. of ferrous sulfate or aluminum sulfate, per 100 sq. ft. of soil. Then test the soil twice a year, and apply more only as indicated. Be sure to wait a year between applications.

Improving soil texture

Nearly all soils are mixtures of sand, silt and clay. Sand particles are the largest. Silt particles are roughly 100 times smaller than those of sand, and clay particles are roughly 1,000 times smaller than those of sand. Clay particles are generally thin flakes, with far more surface area, weight for weight. When they're wetted, the flakes tend to slide but cohere, giving clay soil a characteristic sticky feel.

A soil's ease of gardening is largely determined by its proportions of sand, silt and clay. If a soil is mostly clay, the particles pack together, leaving only tiny pores for air and water. Water clings to the particles, and is held in the pores by capillary action (the same force that wets a wick). As a result, the soil is ill-drained. When wet, it is short on oxygen and inhospitable to roots. (See drawing at right.) If a soil is largely sand, it has many big pores. A little water clings to the particles of sand, but most drains rapidly through the pores. As a result, sandy soils tend to dry quickly and demand frequent watering. If a soil has roughly equal proportions of sand, silt and clay, it also has pores of varying sizes, from small to large. It holds lots of water and air at the same time, so it favors root growth. It also drains just fast enough to prevent waterlogging, even during prolonged wet spells.

Sand, gypsum for heavy soils

Though sand and gypsum (finely-ground calcium sulfate) are often recommended to lighten clay soil, I don't believe either one is practical. Sand works, but you need a lot of it for all but the smallest plots. To really make a difference, you need to mix about 2 in. of sand into the top 6 in. of soil. That's a lot of sand, and a lot of digging. You would need 16 cu. ft. of sand (about six wheelbarrow loads) to spread a 2 in. layer on a 100 sq. ft. plot. Before you try gypsum, ask if your extension service recommends it for your conditions. While it causes clay particles to clump together, opening larger pores in the soil, gypsum also leaches rapidly in areas with abundant rainfall.

Organic matter for heavy or light soils

Organic matter in almost any form—leaves, straw, hay, peat, sawdust, or compost—is the best amendment for both clay soils and sandy soils. Applied regularly, it makes clay soil more open and sandy soil less quick to dry. It also nourishes soil life and improves the soil's capacity to store nutrients.

Organic matter in the soil has a complex fate. It is broken down and digested by microorganisms, earthworms and other soil creatures whose activities release plant nutrients directly and also produce acids that dissolve minerals, releasing more nutrients. The remains and byproducts, an assortment of relatively stable compounds called humus, darken the soil and improve its properties.

Humus has two major effects. First, it holds a lot of water, making sandy soil less quick to dry. (Humus holds up to 90% of its weight in water, while clay holds only 15% to 20%.) Second, humus binds small soil particles together, forming large, stable granules that in turn make the soil more porous, and so improve drainage.

Amending clay soil regularly with organic matter can produce dramatic results. In my last garden, spring and fall, I mulched the asparagus patch with 6 in. of hay. After five years, the top 3 in. of soil had been transformed. It was far darker than nearby unmulched soil. A trowelful would break into big crumbs, not stick to the blade in a lump. Earthworms abounded. The soil was covered with their holes and their castings. The asparagus thrived.

Organic matter can also transform sandy soil. I once visited a south Florida gardener who mulched his entire property with shredded wood and leaves. When he started, the soil was pure white sand. I dug through the mulch

and found that the top few inches of sand were black with humus. Trees, shrubs, vines and groundcovers made a jungle of the property.

Applying soil amendments

How you apply soil amendments depends on how you garden. If you need to alter the pH of the soil around shrubs, trees and other perennials, you must lime or spread sulfur on the surface and mix shallowly to avoid damaging roots. Rainfall percolating through the soil slowly carries the lime or sulfur deeper. To amend the soil texture, you have to mulch and be patient. As my asparagus patch showed, a permanent mulch eventually transforms the soil.

If you are growing annuals, or remaking a vegetable garden each spring, you can be more ambitious. Spade organic matter and lime or sulfur as deep as you wish, within reason, so the zone where most roots grow is amended from the start.

Before you decide you need a soil amendment, talk to your local extension agent. An extension agent helps thousands of gardeners every year and knows the soils of your region. You'll get good advice on the kind and quantities of amendments that suit your garden. □

Mark Kane is an associate editor at Fine Gardening *magazine.*

Regularly adding organic matter to clay soil makes it darker, more granular and more porous, as the top 5 in. of soil demonstrate here. Organic matter also improves the fertility and moisture-retention of sandy soil.

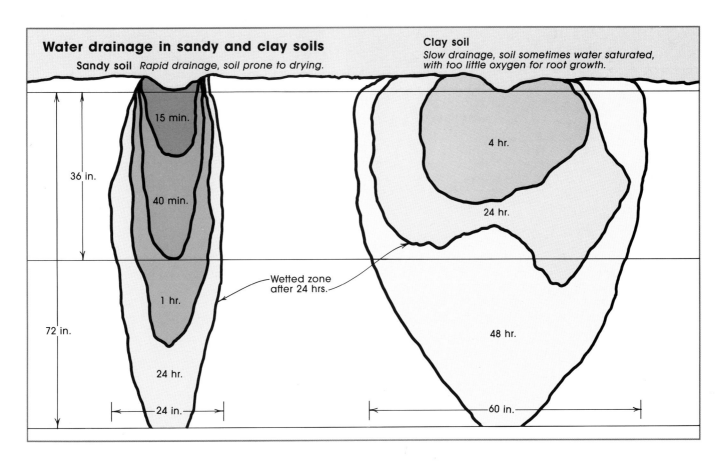

Water drainage in sandy and clay soils

Sandy soil *Rapid drainage, soil prone to drying.*

Clay soil
Slow drainage, soil sometimes water saturated, with too little oxygen for root growth.

15 min.

40 min.

4 hr.

24 hr.

36 in.

Wetted zone
after 24 hrs.

1 hr.

48 hr.

72 in.

24 hr.

24 in.

60 in.

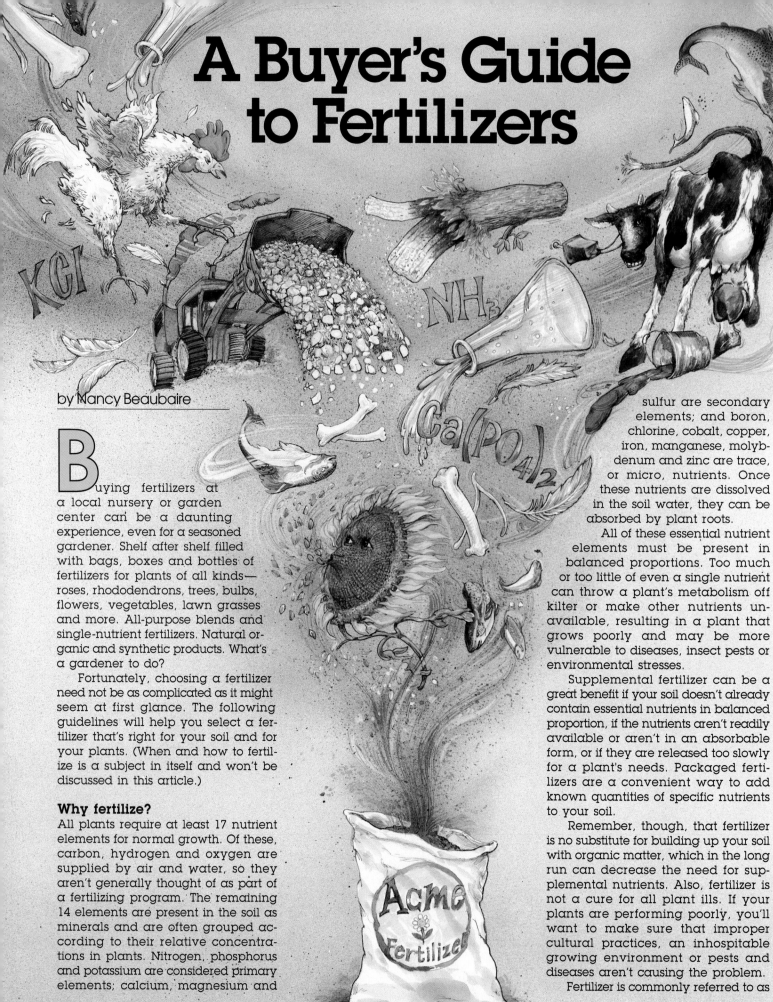

A Buyer's Guide to Fertilizers

by Nancy Beaubaire

Buying fertilizers at a local nursery or garden center can be a daunting experience, even for a seasoned gardener. Shelf after shelf filled with bags, boxes and bottles of fertilizers for plants of all kinds— roses, rhododendrons, trees, bulbs, flowers, vegetables, lawn grasses and more. All-purpose blends and single-nutrient fertilizers. Natural organic and synthetic products. What's a gardener to do?

Fortunately, choosing a fertilizer need not be as complicated as it might seem at first glance. The following guidelines will help you select a fertilizer that's right for your soil and for your plants. (When and how to fertilize is a subject in itself and won't be discussed in this article.)

Why fertilize?

All plants require at least 17 nutrient elements for normal growth. Of these, carbon, hydrogen and oxygen are supplied by air and water, so they aren't generally thought of as part of a fertilizing program. The remaining 14 elements are present in the soil as minerals and are often grouped according to their relative concentrations in plants. Nitrogen, phosphorus and potassium are considered primary elements; calcium, magnesium and sulfur are secondary elements; and boron, chlorine, cobalt, copper, iron, manganese, molybdenum and zinc are trace, or micro, nutrients. Once these nutrients are dissolved in the soil water, they can be absorbed by plant roots.

All of these essential nutrient elements must be present in balanced proportions. Too much or too little of even a single nutrient can throw a plant's metabolism off kilter or make other nutrients unavailable, resulting in a plant that grows poorly and may be more vulnerable to diseases, insect pests or environmental stresses.

Supplemental fertilizer can be a great benefit if your soil doesn't already contain essential nutrients in balanced proportion, if the nutrients aren't readily available or aren't in an absorbable form, or if they are released too slowly for a plant's needs. Packaged fertilizers are a convenient way to add known quantities of specific nutrients to your soil.

Remember, though, that fertilizer is no substitute for building up your soil with organic matter, which in the long run can decrease the need for supplemental nutrients. Also, fertilizer is not a cure for all plant ills. If your plants are performing poorly, you'll want to make sure that improper cultural practices, an inhospitable growing environment or pests and diseases aren't causing the problem.

Fertilizer is commonly referred to as

Illustration: **Gary Williamson**

plant "food," although strictly speaking, the sun is the only source of plant food, or energy, via photosynthesis. Nonetheless, mineral nutrients still play an essential role in the photosynthetic process.

Look before you leap

Before you plunk down your hard-earned dollars for fertilizer, make sure your soil needs it. A soil test is the best way to find out which nutrients are present, and what, if anything, is lacking. Many state university Cooperative Extension Service offices offer inexpensive soil testing (they're listed under government offices in the phone book), or they can refer you to a private lab. You also can test your own soil with a home kit, but the results are likely to be less accurate, and the recommendations are more general. (For a review of these kits, see "Home Soil Testing" on pp. 16-20.)

Soil test results should tell you the existing levels of many of the essential nutrients as well as the soil pH level, a measure of its acidity or alkalinity. The report also will include recommendations for the quantities of each nutrient needed to correct imbalances or deficiencies for the specific plants you're growing. (Nitrogen changes form and quantity rapidly in the soil, so most labs don't test for it, but instead recommend applying it according to the requirements of the specific kinds of plants you're growing.) If the soil pH is too low or too high, you'll need to adjust it to prevent decreased availability of certain nutrients. (For information about how to adjust pH, see "Soil Amendments" on pp. 48-51.)

You also can get a clue about fertilizer needs by looking closely at your plants for characteristic symptoms of nutrient deficiencies or excesses. You'll find descriptions, and sometimes pictures, of these symptoms in many basic gardening books. Even though it's sometimes too late to remedy a nutritional problem the same season you see it, the diagnosis can alert you to the kind of fertilizer you might need to add the next season. It takes experience to interpret symptoms, but observing your plants and their response to fertilizer will, over time, help you recognize imbalances.

Fertilizer choices

Form—Fertilizer is sold in three basic forms: dry granules; solid pellets, tablets or spikes; and liquids or soluble powders. Granular fertilizers are composed of small particles that fall within a specified size range. Broadly speaking, granular fertilizers are the least expensive form per pound of nutrient and are often used for fertilizing lawns and new or established garden areas.

There are both slow- and quick-release granular fertilizers. The particles of a slow- or controlled-release granular fertilizer have been coated with a polymer or other material that reduces the speed with which the nutrients become available. Most of

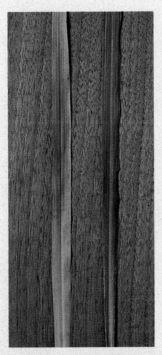

The yellow and purplish-brown color of a leaf from an unfertilized lemon grass plant (left) is symptomatic of multiple nutritional deficiencies. The green leaf (right) was cut from a regularly fertilized plant.

the nutrients of a quick-release granular fertilizer are available soon after application and may last a season or less. (The time can vary depending on the season, climate and soil.)

Slow-release fertilizers are often more expensive, but require less frequent application. They provide a steady supply of nutrients over an extended time and minimize the problems associated with leaching of nutrients or an overdose of certain elements, particularly nitrogen. Usually the label states whether the product is slow- or fast-release.

Solid, pelletized fertilizer is composed of compressed particles that are pre-formed into tiny, bead-like particles, tablets or spikes. Pelletized fertilizers are often more costly per pound of nutrient than granular fertilizers, but many pelletized products are slow-release, which somewhat offsets their

cost. Some gardeners find tablets or spikes, which you place beneath the soil surface, more convenient to apply than granular fertilizer. Pelletized forms are often used for fertilizing container plantings, trees and other individual plants. (See "Fertilizing Trees Makes a Difference" on pp. 58-60.)

Liquid fertilizers are sold as concentrates or powders that must be mixed with water and are then applied directly to the soil or sprayed onto the foliage. Some liquid fertilizers contain only nitrogen, phosphorus and potassium, while others, such as chelated iron, are a source of a single trace element. (Chelation is a process in which a non-nutrient compound is joined to an element, such as iron, to increase its availability.)

Liquid formulations are an excellent source of readily available nutrients and are especially useful for quickly correcting deficiency symptoms and for fertilizing soilless mixes. They also can be used for the same purposes as granular or pelletized fertilizers can, though the liquids tend to be more costly and require more frequent application.

Complete and incomplete fertilizers—All-purpose, or "complete" fertilizers, are so called because they supply the nutrients that plants require in the greatest proportion: nitrogen, phosphorus and potassium. In addition, they may include other essential minerals. If one of these fertilizer blends comes close to the needs of your soil, buying it will be much easier than trying to make up your own mix from individual nutrients.

Some complete fertilizers are sold specifically for certain broad categories of plants, such as flowers, or even more specifically, for a particular kind of plant—roses, for example. In some cases, at least, these tailor-made formulations are based on research trials, and the proportion of nutrients in the fertilizer reflects levels that have been shown to optimize growth of these plants. Some products are geared toward a plant's fertilizer needs as they vary throughout the year. Lawn formulations, for example, often differ for spring and fall applications.

You'll also find fertilizers that provide just one or two nutrients. Blood meal (12-0-0), for example, contains a minimum of 12% nitrogen and no phosphorus or potassium, while bone meal usually contains just phosphorus

(Text continues on p. 55.)

Deciphering a label

You can learn a great deal about a fertilizer from its label. The photos (below and facing page) are side and back labels of the same bag. Nearly all labels include the brand name, fertilizer grade, guaranteed analysis, source of nutrients, net weight and the manufacturer's name and address. Many include information about the purpose of the product and directions for application, as well as cautions about use. (Fertilizers don't pose the same sort of risk that many pesticides do, but granular fertilizers in particular can irritate the lungs, skin and eyes upon contact.) Some manufacturers even offer a money-back guarantee or a toll-free number to call for answers to your questions.

10 LBS. NET WEIGHT

ESPOMA
Plant-tone.
4-3-2

GUARANTEED ANALYSIS

Total Nitrogen (N) 4%
  1% Water Soluble Nitrogen
  3% Water Insoluble Nitrogen

Available Phosphoric Acid
(P_2O_5) 3%
Soluble Potash (K_2O) 2%

Potential basicity equivalent to 80 lbs. of $CaCO_3$ per ton.

THE ESPOMA CO.
MILLVILLE, N.J.

The Natural Organics used in Espoma products are of the finest quality and include Bone Meal, Crab Meal, Animal Tankage, Manures, Cocoa Tankage, Dried Blood, and Cottonseed Meal. Natural materials such as Greensand and Rock Phosphate are also used in Espoma formulations.

Analysis and contents

Net weight—This number is the total weight of the contents of the fertilizer package.

Brand name—This is the specific product name for an individual fertilizer.

Fertilizer grade—The grade is indicated by three numbers on the label, usually located on the front of the package. The numbers refer, in order, to the percentage by net weight of total nitrogen (N), available phosphoric acid (P_2O_5, a form of phosphorus), and soluble potash (K_2O, a form of potassium). For example, a 10-3-4 fertilizer contains 10% nitrogen, 3% available phosphoric acid and 4% soluble potash; a 46-0-0 formulation contains only 46% N.

Guaranteed analysis—In addition to the percentage of primary nutrients, the guaranteed analysis lists the percentage of all other essential nutrients in the fertilizer; these are the minimum percentages guaranteed by state law. The analysis also includes a breakdown of the total nitrogen percentage into the percentage of water-soluble and water-insoluble components, sometimes specifying the exact source of the water-soluble materials, such as urea or ammonium sulfate. (The percentage of water soluble nitrogen tells you how much of this nutrient should be immediately available.)

The remaining percentage of the fertilizer is made up of some combination of chemical impurities, the nonnutritive portion of naturally occurring materials and fillers—substances other than plant nutrients that are added to a mix to provide bulk and prevent caking.

Potential acidity (here "basicity")—Fertilizers that contain nitrogen in the form of ammonium can acidify the soil. The potential acidity is the amount of calcium carbonate that would be required to neutralize the decrease in pH resulting from the application of one ton of this fertilizer.

Derived from—This portion of the label describes the source of each of the essential nutrients in the fertilizer.

(0-10-0), though I've also seen it available in a 1-11-0 formulation.

Do you really need to buy special fertilizers for each kind of plant in your garden? In most cases, probably not, especially if you use a soil test as a guide to the nutrients that are needed. Although it's true that different plants do best with different levels of nutrients, it's nearly impossible to come up with one rose fertilizer, for example, that best suits every soil in the United States. And your roses don't know whether the 5-10-5 formula you applied came out of a box that was labeled for roses or for vegetables (or whether you bought a 10-20-10 formulation and applied it at half the recommended rate).

If you decide your soil requires fertilizer but you haven't done a soil test, you'll need to buy fertilizer based on the manufacturer's suggested uses.

You also could call your local Cooperative Extension Service office and request fertilizer recommendations for the plants that you're growing. In either case, even if the fertilizer you buy isn't exactly right on target for your soil, one application is unlikely to throw the nutrient balance out of whack.

Organic and synthetic fertilizers— Defining what constitutes a natural or organic, versus a synthetic, fertilizer can spark heated controversy among gardeners, but for practical purposes, understanding the differences can be quite simple. Those fertilizers generally thought of as natural or organic are naturally occurring materials derived from plants, animals or minerals that contain elements essential for plant growth. They may be dried, ground, chopped

or physically processed in other ways, but they aren't chemically processed, and they're not combined with any synthetic materials.

Today, the industry trend is to label these products as "natural organic," though sometimes they are sold as "organic." In either case, they include such materials as dried blood, fish emulsion, feather meal, rock phosphate, bone meal, sunflower seed hull ash and seaweed.

Synthetic fertilizers, on the other hand, are produced from raw materials that are changed by a chemical reaction into a form that plants can use. Examples of synthetic fertilizer components include urea, ammonium sulfate, superphosphate, muriate of potash and sulfate of potash. For both synthetic and organic fertilizers, you'll find the sources of the nutrients on their labels.

Compared to synthetics, the complete organic fertilizers usually contain a smaller percentage of nitrogen, phosphorus and potassium. The nutrients are nearly always present in a less soluble form and, consequently, are released more slowly, with the same advantages of slow-release granular fertilizers discussed earlier. (There are exceptions—most of the nitrogen in fish emulsion, an organic fertilizer, is immediately available, while the slow-release synthetic fertilizers act more like a typical slow-release, organic product.)

Naturally occurring materials also contribute some organic matter to the soil, in addition to nutrients. This matter benefits soil microorganisms and, in turn, the soil structure. Some gardeners prefer natural organic fertilizers for environmental or philosophical reasons, as well.

The slower release of nutrients and the lower percentage of nitrgen, phosphorus and potassium in most complete organic fertilizers can be a disadvantage, however, if your soil or the plants you are growing require a quick hit of nutrients or larger quantities of them. Low levels of nitrogen, in particular, are often pointed to as a problem with complete organic fertilizer blends. To get more nitrogen, you might turn to a higher analysis synthetic fertilizer, or supplement the organic fertilizer with a source rich in nitrogen, such as blood meal or urea. □

Nancy Beaubaire is managing editor at Fine Gardening.

Directions

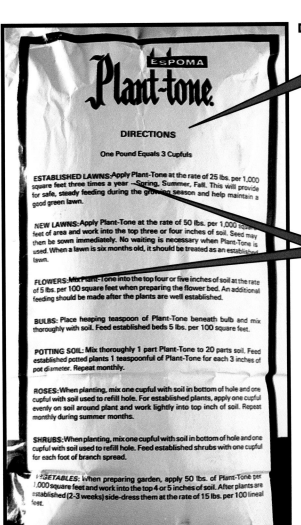

Directions for application—Most fertilizers include recommended application rates and often include suggestions for timing and frequency of application.

Purpose of product— Some, but not all, fertilizers include the manufacturer's suggested use for the product. The information can be as simple as "berry booster," "lawn food" or "improves flower development," or it may include a list of all of the plants that could benefit from this product. Fertilizer labels sometimes provide a detailed explanation of how the product functions in the soil and in plants, as well.

Slow-Release Fertilizers
One application nourishes plants for months

Pellets of slow-release fertilizer pour from a teaspoon into a pot. Made for convenience, the pellets will nourish this impatiens by releasing nutrients throughout the growing season.

by Francis R. Gouin

Slow-release fertilizers are a step toward care-free gardening. They dole out plant nutrients at a measured rate for a predictable time—from one month to two years or more, depending on soil conditions and the fertilizer. Manufactured in pellets, spikes, tablets and pouches, they fertilize plants with one application. (Conventional fertilizers release their nutrients rapidly and are soon depleted, so you must apply them several times during the season.)

Slow-release fertilizers are also less likely than conventional fertilizers to get into groundwaters and surface waters. Applied according to directions, they release their nutrients so slowly that plants absorb up to 90% of them (compared to 50% or less of conventional chemical fertilizers). You can apply conventional fertilizers frequently in small amounts, so plants can absorb nutrients without waste, but you have to be diligent. Their nearly complete absorption makes slow-release fertilizers a good choice for frequently watered containers or fast-draining, sandy soils.

Convenience and safety come at a cost, however. Slow-release fertilizers are generally more expensive than conventional fertilizers, and most of them supply only nitrogen, phosphorus and potassium. Their formulations are similar, ranging from 10-10-10 (10% nitrogen, 10% phosphoric acid and 10% potash) to 18-6-12, so comparing price per ounce gives a rough guide to the cost of their nutrients.

Slow-release fertilizers can be hard to find. Many garden centers carry only one or two. One mail-order source (Mellinger's, Inc., 2310 W. South Range Road, North Lima, OH 44452-9731, 800-321-7444, catalog free) carries four of the fertilizers discussed in this article.

How slow-release fertilizers work
The nutrients in slow-release fertilizers, like those in conventional fertilizers, become available to plants when they dissolve in water. They dissolve faster in moist soil than in dry soil and faster in warm soil than in cool soil.

Slow-release fertilizers use three methods to control how fast their nutrients dissolve. Knowing how they work will help you decide which type suits your plants and soil conditions.

Bacterial release—One type of slow-release fertilizer relies on soil bacteria to make its nutrients available. As soil bacteria digest the coating around

All photos: Susan Kahn

the fertilizer particles, nutrients are released. **Nitroform,** for example, is made from ureaformaldehyde and provides only nitrogen for four to six weeks. Another bacterial-release fertilizer, **Sulfurkote,** is a brand made of nitrogen-rich granules coated with sulfur. Repeated applications of sulfur-coated fertilizers can acidify the soil, but the acidity can be counteracted with applications of lime. Available at garden centers, most bacterial-release fertilizers supply nutrients for three or four months and are used to fertilize lawns.

You can also use bacterial-release fertilizers in containers. If the potting mix is new, inoculate it with bacteria by mixing in about a tablespoon of garden soil or compost before you add the fertilizer.

Weathering release—Some slow-release fertilizers rely on weathering—breaking into fragments—to control how fast they release nutrients. You can buy these weathering-release fertilizers in granules or spikes. **MagAmp,** for example, is a granular fertilizer that is available in fine, medium or coarse particles. The fine particles release nutrients faster than the coarser particles do. Fine particles provide the best results in beds, while the coarse particles last longer and are best for containers. All begin to release nutrients immediately and last through the growing season.

MagAmp contains nitrogen and phosphorus (7-40-6), as well as magnesium. It's well suited for fruiting vegetables, such as tomatoes, as well as for flowering plants. If you use MagAmp, avoid raising soil pH with dolomitic limestone (which also contains magnesium). The two combined can cause magnesium toxicity symptoms such as stunted growth, yellow leaf margins and abnormally small leaves and flowers. Instead, raise soil pH with high-calcium lime.

Jobes' fertilizer spikes also release nutrients by weathering. Drive them into the ground within the root-zone of your plant. The spikes last for one year and have various ratios of nutrients designed to feed trees, shrubs, tomatoes, roses and container plants.

Osmosis release—Some slow-release fertilizers release their nutrients by osmosis—nutrients move from an area of greater concentration to an area of lesser concentration. One type, **Osmocote,** is manufactured in resin-coated pellets. It comes in several analyses that are sold in garden centers, and two, 14-14-14 and 18-6-12, are also available by mail. Two other types, **Gromax** and **Nutri-Pak,** use permeable envelopes. Gromax is 10-10-10 or 12-5-8, and Nutri-Pak is 16-8-8. Osmosis-release fertilizers last from three months to two years or more and are suitable for containers, beds and woody ornamentals.

Osmocote's pellets work best when incorporated into the soil, but they can be applied as a top-dressing. Nutri-Pak packets must be buried in the ground. Gromax envelopes must be buried and are for transplanting or starting plants because they contain moisture-absorbing crystals in addition to fertilizer.

Since osmosis-release fertilizers take two to six weeks to start releasing nutrients, you may want to apply conventional fertilizer until then.

I use all the types of slow-release fertilizers, each for a different purpose. Spikes and envelopes are simple to use in containers when shifting plants into larger pots and for fertilizing individual plants in the landscape. But you must follow package instructions when you use Jobes' spikes in containers. Most spikes require a minimum amount of soil, usually determined by the size of the pot; spikes intended for indoor plants usually require a pot at least 4 in. in diameter. Spikes formulated for tomatoes must be placed at least 6 in. from the plant's stem, requiring a pot at least 12 in. in diameter. Envelopes work best for containers bushel-sized and larger.

The best slow-release fertilizers for lawns are bacterial-release fertilizers because crushing them under foot or with the wheels of lawn mowers won't destroy their release mechanism. ◼

Francis R. Gouin is the acting chairman of the Department of Horticulture at the University of Maryland in College Park, Maryland.

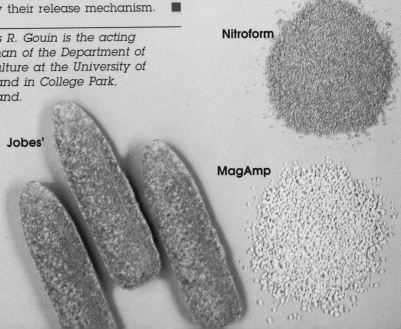

Nutri-Pak

Osmocote

Gromax

Sulfurkote

Nitroform

MagAmp

Jobes'

Fertilizing Trees Makes a Difference

A little goes a long way

by Guy Sternberg

Fertilizing trees with supplemental nutrients is one of the most effective and easiest ways to encourage their healthy growth. Fertilizers are especially important for getting young trees off to a good start; but whether trees are newly-planted or old as the hills, conifers or deciduous species, they can benefit from this boost.

Fertilizer is no substitute for choosing tree species well-adapted to your site, planting healthy stock and providing suitable growing conditions. But, properly applied, fertilizer can help trees better resist and cope with insect pests, diseases and soil problems, especially those trees planted in less-than-optimum conditions.

Why fertilize?

It's a common misconception that fertilizers are "food." Adding fertilizer to the soil does not feed trees. Trees are fueled by sugars produced by their leaves through photosynthesis, but supplemental nutrients, in the form of fertilizer, do provide some of the ingredients needed for photosynthesis.

To grow properly, trees require at least 17 nutrient elements, in balanced proportions, along with suitable conditions of light, moisture, pH, soil and temperature. Of these nutrients, carbon, oxygen and hydrogen can be extracted directly, as needed, from air and water, and don't need to be supplemented outdoors. Nitrogen is also present in the air, but must be combined with other elements in the soil into a form that can be absorbed. The remaining nutrients are present as soil minerals. (See nutrient chart on p. 59)

These mineral nutrients are frequently categorized as primary, secondary and trace, based on their relative concentrations, but all are critical for proper tree growth. The primary nutrients are nitrogen, phosphorus and potassium; the secondary ones are magnesium, calcium and sulfur;

Fertilizing trees is an easy and inexpensive way to support their growth. The yellow needles of the cypress tree shown above are symptomatic of a nutrient deficiency. In this instance, the tree lacked iron, a problem that was easily remedied once the tree was fertilized, below.

and the trace, or micro, nutrients, include iron, chlorine, manganese, zinc, boron, copper, molybdenum and cobalt.

Do your trees need fertilizer?

Soil tests—Initially, a soil test is mandatory to determine whether and how much to fertilize your trees. (State university Cooperative Extension Service offices offer inexpensive soil testing or can recommend a private lab.) The results of a complete test should include existing levels of all the required nutrients, except nitrogen, and in most cases, recommendations for quantities needed to correct deficiencies or imbalances. Labs usually don't test for nitrogen because it is so mobile, and the common practice is to fertilize with it regularly.

The soil pH level, a measure of alkalinity or acidity of the soil, is usually tested for as well, and should be adjusted if it is too high or too low. Otherwise, the availability of certain nutrients will be limited, even if they are present in the soil. Conifers and most acid-loving broadleaf trees prefer a pH of 5.0 to 6.0, while most deciduous trees do best at a pH of about 6.5. A few, such as Chinquapin oak, blue ash and hackberry tolerate much higher pH levels.

Deficiency symptoms—Soil test recommendations are the primary guide for fertilizing your trees, but you also should watch for symptoms of poor nutrition. Check for overall reduced growth, sparse or small foliage, abnormal leaf color and, on deciduous trees, early onset of fall color.

If visual symptoms appear to conflict with your soil test, you can send leaf samples to a specialty lab for nutrient analysis, which will indicate whether the nutrients present in the soil have, in fact, been taken up by the tree. Leaf analysis should rarely be necessary in a home landscape situation, but could be a worthwhile investment where the health of a prized tree is jeopardized. Contact a certified, professional arborist for lab recommendations.

Choosing a fertilizer

Formulations—Most all-purpose tree fer-

Photos: Eugene B. Himelick

tilizers contain a mixture of nutrients. The percentage, by weight, of nitrogen, phosphorus and potassium are indicated, in that order, by three numbers on the analysis label. For example, a 10-6-4 fertilizer contains 10% nitrogen, 6% phosphorus and 4% potassium. The label may also list percentages of other nutrients, as well as properties such as residual acidity (calcium carbonate equivalent) and salt-index (burn potential). Single-nutrient fertilizers are available, as well.

Tree fertilizer is most commonly sold in several forms—quick-release granular, slow-release tablets and liquid concentrate. Granular fertilizers are the least expensive, and most of their nutrients are immediately available. They are effective for about one season, depending on soil conditions and climate. Slow-release tablets contain nutrients that gradually dissolve over one to three seasons. They are often more expensive than granular fertilizers, but less subject to leaching. Liquid formulations, which are usually sold as soluble crystals, contain the most readily available nutrients. Other forms are available, as well, including coated, slow-release pellets, organic by-products and tree spikes, which are pre-formed so they can be easily inserted into undisturbed soil.

How much to fertilize—Different tree species, and the same species growing under different conditions, can vary in their fertilizer requirements. For example, conifers generally need lower nutrient concentrations than broadleaf trees. But as a rule of thumb, most trees growing in average soil respond well to an annual application of nitrogen, which is very labile, and equal or

Too much fertilizer, particularly nitrogen, can damage trees, especially if it is applied to dry soil. Here, the author examines a young pine whose needles were "burned" when, under drought conditions, the nitrogen pulled moisture out of the roots, dessicating the foliage.

lesser amounts of phosphorus and potassium. Recommendations range from 2 lbs. to 6 lbs. of nitrogen per 1000 sq. ft. of root-zone soil, the area where the bulk of the absorbing roots are concentrated. (See Computing Quantities on p. 60.)

The National Arborist Association recommends 3 lbs. of nitrogen per 1000 sq. ft., for average conditions, but I prefer to use a lower dose—about 2½ lbs. This amount is adequate for steady growth in my soil, and minimizes the risks of an overdose. Too much nitrate form of nitrogen can "burn," or dessicate, the tree roots and foliage, damage underlying turf and stimulate excessive succulent growth, which invites disease problems, winterkill and wind damage. And, an excess of certain nutrients can make others unavailable.

I hedge my bets with a cautious fertilizer program, which you can use as a guide for developing your own. To get my trees off to a good start, I add slow-release tablets, which include micronutrients, at planting time at the rate recommended on the label. Each subsequent year, I spread an all-purpose, granular fertilizer on the soil surface. If some trees aren't growing vigorously or look deficient, I also may reapply slow-release tablets after two years. (I don't count them as part of the 2½ lb. rate.) For trees that are obviously nitrogen-deficient, I sometimes increase the nitrogen dose to no more than 6 lbs. per 1000 sq. ft., preferably split into spring and fall applications. If you use a high dosage, make sure you compute the root-zone area precisely and ask your supplier for a fertilizer with a large percentage of W.I.N. (water-insoluble nitrogen), or a low-salt index, which will avoid overdose problems.

Because it's expensive and short-lived, I use liquid fertilizer only to temporarily correct specific micronutrient problems of container-grown trees and as a shot in the arm for new transplants. I spray the leaves of my new trees once or twice a month during spring and early summer, using about one-half of the recommended dose.

Fertilizing methods
Any fertilizer should be placed close to, or on, the soil surface, where it will be absorbed by the fibrous, actively-growing roots of the tree, most of which are located no deeper than about 1½ ft. That said, each form of fertilizer is best-suited to a particular method of application.

Broadcast—Granular fertilizer should be spread uniformly, or broadcast, over the root-zone of established trees and watered in. I spread it with a cyclone, or whirlybird, spreader—a small, hand-held device that allows the granules to flow out of the bottom. (You also can scatter granular fertilizer by hand like chicken feed, but this method is not as precise.) To

Nutrients and their Concentration in Tree Foliage

Element (Symbol)	% of Foliage	Deficiency Symptoms
Nitrogen (N)	1.5	Pale young leaves; chlorosis (yellowing) and premature abscission of older leaves; small, reddish twigs.
Phosphorus (P)	0.2	Small, dark leaves with purple veins; severe stunting.
Potassium (K)	1.0	Interveinal chlorosis of older leaves; scorch and marginal necrosis (browning and death).
Calcium (Ca)	0.5	Chlorosis and scorch of young leaves; tip and terminal bud dieback; stunted roots.
Magnesium (Mg)	0.2	Marginal chlorosis of older leaves; interveinal necrosis at leaf center; undersized fruit.
Sulfur (S)	0.1	Overall uniform chlorosis and stunting of young leaves.
Iron (Fe)	0.01	Sharply defined interveinal chlorosis of young leaves; branch dieback.
Chlorine (Cl)	0.01	General reduction in tree growth.
Manganese (Mn)	0.01	Marginal interveinal chlorosis of older leaves; interveinal necrosis.
Zinc (Zn)	trace	Mottled and striped chlorosis of older leaves; leaves form dwarfed rosettes.
Boron (B)	trace	Terminal shoot dieback; sparse foliage on new lateral shoots; brittle, red-veined, distorted leaves.
Copper (Cu)	trace	Veinal chlorosis; terminal leaves form rosettes, terminal growth dieback;
Molybdenum (Mo)	trace	Marginal chlorosis of old and young leaves followed by interveinal chlorosis; leaf cupping.
Cobalt (Co)	trace	Leguminous trees will not develop nitrogen-fixing bacterial nodules.

Note: These symptoms may vary according to the species, the season and the severity of the deficiency. Reduction in growth may precede the appearance of deficiency symptoms.

ensure even distribution, I walk in over-lapping circles around the root-zone as I spread the fertilizer. The only side effect of this method is that you can get dizzy if you don't look up!

Subsurface—Slow-release fertilizers, which are sold as tablets or briquettes, should be placed deep enough under the soil surface to stay moist throughout the growing season, but no deeper than 1 ft. Most types eventually dissolve fully; the briquettes don't, but their nutrients will still become available.

To fertilize a newly-planted tree, I place several tablets around, not below, the rootball, about 10 in. apart, or as di-rected by the label. For an established tree, I make 8-in.-deep holes at 2-ft. inter-vals, just outside the drip line, drop two or three tablets in each hole and refill the holes. I prefer making the holes with an auger, which doesn't compact the soil and leaves a nice pile of loose soil for refilling the hole, but you can use a pointed bar, stout stick or a trowel to poke holes, too.

Spray and soil-injected—Liquid formula-tions can be sprayed on the foliage, or poured on or injected ("fertigated") into the soil. If you use a sprayer, make sure the liquid is diluted accurately—some sprayers automatically dilute it; others re-quire that you do so beforehand. Adding a few drops of a compatible surfactant such as dish detergent to the spray will decrease its surface tension and improve its absorption. Allow the runoff to soak into the ground.

If you use a soil injector—which you can purchase at a garden center—or if you pour the liquid on the soil, apply 200 gal. of solution, mixed according to label directions, per 1000 sq. ft. of root-zone. Try to get uniform coverage. Fertilizer should be injected about every 3 ft. until the solu-tion is completely used up.

Timing of application

Fertilizer is most effective when it's ap-plied during peak growth periods. For soil applications, fertilize from late winter through early spring, when root growth is most active, or during a second root growth spurt, from early fall until the soil becomes either too dry or colder than 40° F. Spray on foliar fertilizers from late spring through early summer, before the maturing leaves have a chance to be-come tough or waxy.

Here in U.S.D.A. Zone 5, and in other si-milar cold winter/warm summer conti-nental climates, roots grow most actively from about mid-March to early May, and then again from about mid-September to early November. Leaf growth is most ac-tive from about early May through mid-July. In climates with milder winters and drier summers, such as those typical of

The author fertilizes a cherry tree in early spring (above), when root growth is active and rainfall enables the fertilizer to be absorbed easily. Granular fertilizer can be applied efficiently with a whirlybird spreader (below), which releases the granules in a steady flow.

the West Coast, root growth is generally most active during the rainy season.

I prefer to apply soil fertilizers in the early spring. At that time, there's usually enough rain to supply moisture for nutri-ent absorption, and my winter tree chores are completed.

If you have ample irrigation water or fall rains, you might want to fertilize in the fall, when the still-warm soil enhances nu-trient absorption. Better yet, split your an-nual fertilizing rate into two doses, timing them with the opening of spring cherry blossoms and the twilight of fall colors.

Remember that trees must expend metabolic energy to absorb and translo-cate the nutrients required to generate still more energy. Those that have been pam-pered with a sustained nutrition program will do better than those that are stressed or living a feast-or-famine existence. □

Guy Sternberg manages Starhill Forest in Petersburg, Illinois, and works with the Illinois Department of Conservation.

Computing quantities

To figure out how much granular or liquid fertilizer to apply, first measure the area of the root zone, the surface area of soil that extends from the trunk to 6 ft. to 8 ft. beyond the crown spread, or drip line, for large or mature trees, and 1 ft. to 2 ft. beyond for young trees. Even though some of the roots of established trees extend out farther, fertilizing this portion of the root zone should be enough to sustain healthy growth.

The root zone area is generally visualized as circular, but it's easier to calculate its area as a square or a rectangle. To do this, draw an imaginary square around the outer perimeter of the drip line and calculate the number of square feet within this boxed area. This is the approximate root zone area. (If the root zone overlaps a paved surface, subtract that square footage from the total.) Roots of trees with a narrow, upright shape extend out much farther than the drip line, so I calculate their root zone to equal that of a tree with an equivalent trunk diameter but a broader canopy.

Next, calculate the total number of pounds of nitrogen required to fertilize this root-zone area. To do this, divide the square footage of the root zone by 1000, and multiply the result by the desired nitrogen rate (2 lbs. to 6 lbs. per 1000 sq. ft.). Finally, to figure out how many pounds of a particular fertilizer to use, divide the total pounds of nitrogen required (the number you just calculated) by the percentage of nitrogen in the fertilizer, used as a whole number. Then multiply the result by 100.

Here's an example. Let's assume you want to fertilize 100 sq. ft. of root zone at a rate of 2 lbs. of nitrogen per 1000 sq. ft., using a 15-7-6 fertilizer. You will need 0.2 lbs. of nitrogen to cover this area, which will require about 1.3 lbs. of fertilizer. It's usually best to use a complete fertilizer with nitrogen and equal or lesser amounts of phosphorus and potassium, unless your soil test report indicates otherwise.

Seaweed Comes Ashore

A natural soil amendment and plant growth stimulant

by Delilah Smittle

Seaweed is good for the garden. Mixed in the soil, it slowly releases nutrients that plants need, while improving soil texture. Since it is particularly rich in micronutrients such as iron, copper, zinc, boron and manganese, seaweed offers a natural remedy for soil with a micronutrient deficiency. Seaweed also contains large quantities of hormones that stimulate plant growth. Plants in seaweed-amended soil grow faster and larger than plants in soil with a comparable amount of conventional fertilizer.

A traditional soil amendment in coastal gardens, seaweed is now formulated in extracts and granular products that you can find on garden center shelves and in catalogs of garden suppliers (see Sources on p. 62). Fresh seaweed and dried granular seaweed must break down in the soil to release their nutrients and hormones. A foliar spray of seaweed extract and water makes the nutrients and hormones available to plants faster. Research has shown that plant health can improve within days after the spray is applied. Foliar seaweed sprays rapidly correct nutrient deficiencies, improve fruit set and help a plant endure environmental stress, including drought and frost.

Where it started

Coastal gardeners have long collected seaweed and composted it or used it fresh as a mulch in their gardens. In the British Isles, 19th-century gardeners grew potatoes of superior flavor in layers of sand and seaweed on bedrock. Traditionally, seaweed is raked from the sea by hand, piled into skiffs

Seaweed has been used for centuries in coastal gardens as mulch or compost to provide minerals and nutrients. Today, rockweed is still harvested by hand off the coast of Nova Scotia, but it will be manufactured into seaweed extract and meal and made available to gardeners everywhere.

> *Seaweed sprays rapidly correct nutrient deficiencies, improve fruit set and help plants endure environmental stress.*

and brought to shore. It is time-consuming, heavy work. A small boatload of fresh seaweed weighs 4,000 lb. to 5,000 lb. Not surprisingly, the discovery of synthetic fertilizers in this century eclipsed labor-intensive and slow-acting organic amendments, seaweed among them.

Seaweed's emergence as a tonic for plants began with British experiments with seaweed as a replacement for hemp during World War II. Scientists learned that as a rope substitute, seaweed was hopeless because it dissolved in water. This discovery, however, led to a process for liquidizing and concentrating seaweed, making it possible to bottle and to transport economically its minerals and hormones. Drying seaweed over low heat led to the production of seaweed meal, a source of minerals and vitamins for livestock feed, and a concentrated soil amendment. Today, gardeners can readily find seaweed extract and seaweed meal.

The primal supermarket

Seaweed is a rootless plant in the *Fucus* family that floats freely or clings to rocks by holdfasts (root-like or disc-shaped plant parts that attach seaweed to rocks but don't absorb nutrients). Seaweed photosynthesizes the sunlight that reaches it through shallow water, and it absorbs nutrients

Photo: Percy Cottreau

Eelgrass, a seaweed with very long, narrow leaves, grows abundantly along the North Atlantic Coast. This wheelbarrowful was collected in Falmouth, Massachusetts, and dried for use as mulch.

from seawater through its leaves. Since the ocean receives runoff from the entire earth, it contains all known minerals, trace elements and vitamins. This primal supermarket supplies a more complete diet for sea plants than any plot of rich soil or fertilizer provides for land plants. Seaweed contains 60 or more minerals and several plant hormones. It is not, however, a complete fertilizer. It has a fair amount of nitrogen and potash, but very little phosphorus, a major plant nutrient.

Only a few seaweeds are harvested commercially. Norwegian kelp (*Ascophyllum nodosum*), a brown algae, is the seaweed most used in gardening. Norwegian kelp is gathered off the coasts of England, Ireland and Norway and both the Atlantic and Pacific coasts of North America where it is called rockweed. Gulfweed (*Sargassum*), a floating sea plant, is harvested off the coast of North Carolina. Giant kelp (*Macrocystis*) is collected in the Pacific Northwest.

How seaweed enhances plant growth

Seaweed is constantly worn down by tides and eaten by fish, so it must grow rapidly to survive. Studies at the University of California showed that a frond of seaweed can grow a foot a day, given optimal conditions. The same growth hormones that prompt such rapid growth in seaweed, when applied to plants as a foliar spray, can increase the speed of cell divi-

SOURCES

There are many mail-order catalogs that list one or more seaweed products. Below are garden product companies that carry several seaweed products.

Age-Old Organics, P.O. Box 1556, Boulder, CO 80306. 303-499-0201. Catalog free.

Bramen Co., Inc., P.O. Box 70, Salem, MA 01970-0070. 508-745-7765. Catalog free.

ENP Inc., Box 218, 603 14th St., Mendota, IL 61342. 800-255-4906. Catalog free.

Integrated Fertility Management, 333 Ohme Gardens Rd., Wenatchee, WA 98801. 800-332-3179. Catalog free.

Mellinger's Inc., 2310 W. South Range Rd., North Lima, OH 44452. 800-321-7444. Catalog free.

The Natural Gardening Co., 217 San Anselmo Ave., San Anselmo, CA 94960. 415-456-5060. Catalog free.

Necessary Trading Co., P.O. Box 305, New Castle, VA 24127. 800-447-5354.

Nitron Industries, Inc., P.O. Box 1447, Fayetteville, AR 72702. 501-750-1777. Catalog free.

North American Kelp, 41 Cross Street, Waldoboro, ME 04572. 207-832-7506. Catalog free.

Peaceful Valley Farm Supply, P.O. Box 2209, 110 Springhill Blvd., Grass Valley, CA 95945. 916-265-4769. Catalog $2.00

SeaSpray Products, Inc., P.O. Box 99432, Seattle, WA 98199, 206-282-2850. Catalog free.

sion and elongation in those plants. The hormones also increase root growth when applied to the soil as meal, or when a seaweed extract is used as a root dip.

In recent turf tests at Virginia Polytechnic Institute in Blacksburg, plots sprayed with seaweed extract had 67% to 175% more roots than untreated plots. Plots treated in fall showed a 38% increase in spring growth over untreated plots and showed 52% more roots.

In tests at South Carolina's Clemson University, seeds soaked in liquid seaweed extract showed rapid germination, and the resulting seedlings had increased root mass and stronger plant growth than seedlings from untreated seeds. They also had a higher survival rate. Soaking plant roots in seaweed extract reduces transplant shock and speeds root growth. Seaweed foliar sprays promote faster, stronger stem and leaf growth, and earlier blossoming and fruit set when sprayed on leaves and flower buds.

Seaweed as fertilizer

Seaweed improves soil fertility in several ways. Seaweed's nutrients and hormones are directly available to plants. Mannitol, a compound found in seaweed, enables plants to better absorb nutrients from the soil. The rapid breakdown of carbohydrates in seaweed stimulates beneficial soil bacteria that fix nitrogen and make it available to plant roots. These activities reduce the need for chemical fertilizers, and when seaweed is used with them, enhance their effects.

Fresh seaweed should be rinsed free of salt before it is used in the garden. Mark Schonbeck, former researcher at the New Alchemy Institute, East Falmouth, Massachusetts, recommends applying fresh seaweed to soil only once every four or five years, because seaweed's high salt and boron content can poison the soil. Seaweed meal can be used in place of fresh seaweed, but at a greatly reduced rate. T.L. Senn, a long-time seaweed researcher, advises using 250 lb. to 500 lb. of seaweed meal per acre, or having it make up 3% to 5% of a homemade fertilizer mixture.

A balanced organic fertilizer can be created by mixing fresh seaweed or seaweed meal with manure or fish meal, both of which supply sufficient phosphorus. Seaweed is also a good soil conditioner, and can add as much humus to the soil as manure can.

Robert Kourik, an organic gardening specialist, suggests using 1 lb. of seaweed meal per 100 sq. ft. of soil, or ¼ tablespoon of liquid concentrate to 1 gal. of water for a foliar spray in intensive vegetable gardens. No matter what formulations you use—fresh, dried or liquid—don't exceed the recommended quantities, because excessive amounts of seaweed can stunt plant growth rather than encourage it.

Seaweed as pest control

Some scientists believe that seaweed has developed antitoxins to fend off bacteria and viruses in the ocean. In the garden, these antitoxins interrupt the reproductive cycles of some insects, and appear to repel others. Seaweed also reduces fungi when applied to plants or soil. In tests at the University of Maryland, seaweed meal reduced soil nematodes in turf grass plots. Clemson University studies showed fewer aphids and flea beetles on foliar-treated plants, and other studies showed resistance to spider mites and scab. In Clemson studies, fruits and vegetables treated with seaweed didn't grow mold and thus had a longer shelf life.

Using seaweed

You can apply seaweed as a mulch or as a soil additive, or incorporate it in a compost pile (its ability to activate soil bacteria makes seaweed an excellent compost starter). But the preferred method of application is as a foliar feed.

Clemson University tests compare growth of tomato plants watered with plain water (pots 1, 7-9) to plants watered with seaweed extract diluted 1:100 (pots 2-6), and seaweed extract diluted 1:50 (pots 10-12).

Seaweed has antitoxins to fend off bacteria and viruses and to repel insects in the garden.

For a head start on the growing season, you might want to presoak seeds in diluted seaweed extract for 20 minutes before planting. Then water the seedlings regularly with the same solution until strong growth appears.

Apply seaweed meal to the soil as soon as the ground can be worked in spring, because the meal needs time to break down. Work the meal into perennial beds when the plants break dormancy.

Apply foliar sprays once or twice a month during the growing season. Spraying in late fall supplies phosphorus and zinc to plant roots, and increases the frost tolerance of grass, vegetables and perennials. A late-season foliar treatment can yield a longer harvest of vegetables.

It's best to buy small quantities of seaweed products and to store them in a cool, dark place. They have no preservatives and will degrade after opening, though probably not before the end of the gardening season. Seaweed meal keeps longer than extract—meal will keep for a year, but can develop mold if stored in a damp place. A quart to a gallon of seaweed concentrate, or a 40-lb. bag of seaweed meal, should be enough to treat an average city lot for a season, at a cost of less than $20. □

In other Clemson tests, tomato, bean and corn plants treated with seaweed extract showed resistance to insects, disease and drought. They also produced meatier vegetables with longer keeping qualities than untreated plants did.

Delilah Smittle is an assistant editor at Fine Gardening *magazine.*

Photos: Clemson University, Clemson, South Carolina

A Mulch Primer

Year-round cover aids plants and soil

A mulch of shredded cedar bark makes a dark rectangle in a bed of perennials. The new mulch, 3 in. deep, came from one 3-cu.-ft. bag. Mulch offers a variety of benefits to plants, soils and gardeners.

Photo: Mark Kane

by John W. Mastalerz

Mulch is as much a part of gardening as soil and fertilizer, planting and pruning. Almost every gardening book and magazine recommends its use. But just what is mulch, and what role does it play in the garden?

Over the course of 30 years of teaching and research at the Pennsylvania State University, I learned a good deal about mulch and how it works, about the pros and cons of common mulching materials, and about how to apply them in a garden. With this information, you can decide if mulching is right for your plants and which mulch suits your garden.

The benefits of mulch

A mulch is a material applied to the surface of the soil that conserves soil moisture, moderates soil temperature, controls weeds, maintains the physical structure of the soil, and prevents soil erosion—just like leaf litter in a forest. Almost anything that you can spread over the soil will serve as a mulch. It can be organic—leaves, pine needles, bark chips—or inorganic—stones, chips of marble, plastic film or even aluminum foil. I'll talk mainly about organic mulches, since they improve soil and inorganic mulches do not.

Mulch slows the loss of soil moisture by blocking sunlight and by insulating the soil. When sunlight strikes unprotected soil, the surface heats up, speeding evaporation. Mulch keeps the temperature of the soil from rising above that of the air, thus limiting the amount of water lost through evaporation. A layer of mulch also conserves soil moisture by shielding the soil surface from the wind and from rising warm-air currents; both carry evaporating moisture away from the soil.

The ability of mulch to moderate the temperature of the soil has other benefits. Under an insulating blanket of mulch, soil temperatures are relatively steady, buffered from large, daily swings in air temperatures. Most plants grow better in a soil that has a fairly constant temperature. Extremes of heat or cold halt root growth.

In cold-winter climates, steady soil temperatures can be critical to the survival of some plants, especially herbaceous perennials. Bare, frozen soils may absorb enough solar energy during the daylight hours for the surface layers to thaw. As the air temperature drops after sundown, the soil refreezes. Soil expands as it freezes, sometimes pushing plant roots up out of the ground. This heaving of the soil (not freezing or thawing) damages unprotected plants. By excluding sunlight and by insulating the soil, mulch prevents the freezing and thawing of the upper layers of the soil. Gardeners in cold climates can prevent heaving by mulching deeply with loose materials such as pine needles or evergreen boughs when the ground freezes. (You should remove most or all of the mulch as new growth begins in the spring to let the perennials emerge.)

A mulch can save a gardener a lot of weeding. Many weed seeds need light to germinate, and mulch prevents light from reaching the soil. Those weed seeds that do germinate are usually easy to pull because the struggle to push through the mulch makes them spindly. Weeds that germinate on top of the mulch are also easy to pull, if you get to them before their roots reach the soil. Don't use mulch to smother established weeds. Many are strong enough to push back through it. Unless you apply an unusually thick layer of mulch, pull or kill actively growing weeds first, then apply the mulch.

Over time, organic mulches improve the fertility and the physical structure of the soils they cover. As mulch particles in contact with the soil decay, they release the nutrients they hold. They also release chemical compounds that help individual soil particles stick together in aggregates that are water-stable (not easily broken apart when wet). A soil with a high level of aggregation drains well and allows oxygen, which is vital to root growth, to flow into the pores between the soil aggregates.

Mulch helps prevent soil erosion. Rain and sprinkler droplets hit unprotected soil forcefully enough to shatter soil aggregates and dislodge individual soil particles. Once dislodged, soil particles can wash away on even the slightest slope. A mulch breaks the fall of droplets and keeps the soil intact. It also prevents mud from spattering up onto plant leaves, foundation walls and walks. Finally, mulch slows the flow of water on slopes so that the water has time to soak into the soil.

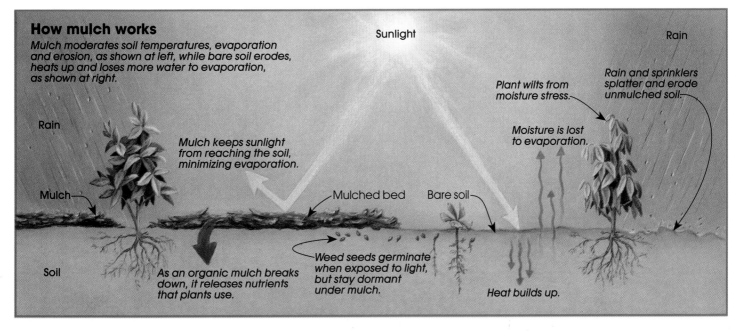

How mulch works

Mulch moderates soil temperatures, evaporation and erosion, as shown at left, while bare soil erodes, heats up and loses more water to evaporation, as shown at right.

Sunlight

Rain

Rain

Plant wilts from moisture stress.

Rain and sprinklers splatter and erode unmulched soil.

Moisture is lost to evaporation.

Mulch keeps sunlight from reaching the soil, minimizing evaporation.

Mulch

Mulched bed

Bare soil

As an organic mulch breaks down, it releases nutrients that plants use.

Weed seeds germinate when exposed to light, but stay dormant under mulch.

Soil

Heat builds up.

Illustration: Steve Buchanan

What you'll find at the garden center—Most garden centers offer a selection of barks and stones in plastic bags. As a service to customers, the garden centers sometimes open bags to display their contents and label them with quantity and price.

Mulch can have one more benefit: many mulching materials improve the look of the landscape, particularly a newly planted one. The natural brown of shredded leaves or the gray of weathered wood chips shows plants off to advantage. The texture of the mulch can also be attractive. Some gardeners prefer bare soil as a background for plants, but I find that unless soil is frequently cultivated, it tends to pack down and look like a parking lot.

Drawbacks of mulching

Mulches aren't perfect. While they provide many benefits, mulches may create problems. Mulches control weeds very effectively, yet some mulches, such as straw, hay or manure, can be loaded with weed seeds and can add to your weed problems rather than reducing them.

Mulch can also provide a favorable environment for animal pests. Slugs find mulches a cool, moist hideout during the daylight hours. Mice, voles and other rodents like the cover mulch provides. They tunnel under many mulches, then feed on the stems and roots of your plants.

Some mulches can present a fire hazard. Straw, hay, peat moss and pine needles ignite readily and burn rapidly. In places where smokers might drop lighted cigarettes, skip these flammable mulches and use bark, wood chips or stones instead.

Some researchers suggest that mulches high in carbon and low in nitrogen (woody materials, particularly sawdust) can cause a temporary nitrogen deficiency in the soil, which limits the growth of plants. Microorganisms at the soil surface that use the carbon compounds in mulch as a source of energy also need nitrogen. If the mulch is low in nitrogen, they take it from the soil, depriving the plants. If you use sawdust and your soil is low in nitrogen, scratch a sprinkling of nitrogen fertilizer into the soil before mulching.

Mulching materials

Because mulch can do so many different things in the garden, it seems to have almost magical properties. But most mulches are common, ordinary materials. Many are crop residues or food by-products, often generated in such volume that they create a disposal problem. Using them as mulch

is a way of recycling them. (For a close look at materials commonly used as mulches, see the photos on the facing page.)

What are the characteristics of a good mulch? I look for materials made up of relatively coarse particles that stay loose. Fine particles tend to pack down, allowing the mulch to soak up water from the soil and transmit it to the surface where it evaporates. The spaces between coarse particles are too large to pull water from the soil. The dead air space between coarse particles also insulates the soil, moderates soil temperature and allows oxygen to reach the soil. Materials that pack transmit heat to and from the soil surface more readily than looser materials and inhibit the exchange of gases.

The availability of mulches varies from region to region. Bags of bark chips seem to be sold almost everywhere, but you're likely to find salt marsh hay only near the ocean and cottonseed hulls near cotton farms. To get an idea of what is readily available in your area, see what fellow gardeners are using, visit local garden centers and look in the classified ads. You may find a mulch that you can

Photo: Chris Curless

get for free. If you decide to buy mulch, shop around. Because mulch is bulky, transportation and handling charges may account for most of its cost, not the actual value of the material. In general, you'll do better buying in bulk.

Every mulch has its pros and cons. If you talk to fellow gardeners, you're likely to find that they have strong preferences. Pine needles are my first choice. They're attractive and free from weed seeds. They don't pack down, but they stay in place. They are also easy to spread between and under plants of all sizes. Some gardeners fear that pine needles lower soil pH as they break down, but they decay so slowly that I don't worry about any dramatic changes in pH. Pine needles can be a fire hazard, but I find that in my yard, even if they dry out on the surface, they stay wet underneath. They're expensive to buy where I live, but if you're willing to gather them yourself, as I do along the roadside or from lawns in the neighborhood, you can get them at no cost.

Applying mulch

Unless you're trying to cover a steep incline, mulching is easy. Start by carrying the mulch to the area where you want it. A wheelbarrow or a garden cart is the least strenuous means of transporting mulch, which can be very heavy when wet. Pour the mulch from the bag or scoop it from the wheelbarrow into piles between plants. Then spread it to cover the area. I like to get down on my knees to spread mulch. You don't have to do a perfect job, but the more uniform the mulch's depth, the more attractive and effective the mulch will be.

The depth to which you mulch depends on the material you are using. Loose mulches composed of large particles should be laid on more thickly than finer mulches. You'll have to experiment to find the best depth for a given material in your garden, but as a rule of thumb, I recommend 1 in. to 2 in. of fine materials such as sawdust or buckwheat hulls, 2 in. to 3 in. of coarser materials such as bark, 3 in. to 4 in. of pine needles, and as much as 6 in. of loose hay. You *can* give your garden too much of a good thing, though. Mulch laid too thickly can prevent water and oxygen from reaching plant roots.

Avoid packing mulch against plant stems. Mulch in close contact with stems creates conditions that favor the development of diseases capable

A close look at widely available mulches

Bark
- Chips screened to uniform size; sometimes ground or shredded; light brown to reddish brown in color.
- Available in bags from garden centers; locally available in bulk.
- Spread 2 in. to 3 in. deep.
- Durable, slow to decay, easy to spread.

Wood chips
- Hardwood branches run through a chipper.
- Locally available; tree services and power companies may have them for sale; you can make your own with a chipper.
- Spread 2 in. to 3 in. deep.

Leaves
- Spread 2 in. to 3 in. deep.
- Tend to pack when whole; best shredded with lawn mower or shredder.

Pine needles
- Spread 3 in. to 4 in. deep.
- Slow to decay; possible fire hazard; expensive unless you collect your own.

of injuring or even killing some plants. Hold the mulch away from the stem or trunk as you mulch.

Organic mulches require occasional replenishment as they break down. Leaf mold (decomposed leaves), grass clippings or shredded leaves may have to be replenished more than once a year. More durable materials, such as pine needles, wood chips or bark, can go two or more years without a fresh layer. Your climate will also determine how often you need to add mulch. In the South, where heat and humidity speed decomposition, materials that are durable in colder climates may be ephemeral. And no matter where you are or what kind of mulch you use, digging inevitably causes mulch to disappear into the soil. □

John W. Mastalerz is a professor emeritus of horticulture at the Pennsylvania State University. He mulches his garden in South Chatham, Massachusetts.

Mulch, Don't Dig

Layers of newspaper prepare a new garden bed

Author Carney starts new ornamental plantings by spreading layers of newspaper and mulch to smother the existing plants. Here she's converting a portion of lawn to a bulb garden. It's early spring, and the newspaper and mulch
are freshly laid. The grass underneath is still green, but will die for lack of sunlight by the end of the growing season. Carney cuts out a strip of sod to
mark the new garden's border and make mowing the adjacent lawn easier.

by Nancy Carney

The hardest part of lay-
ing out a new flower or shrub
garden is getting started.
Dreaming about plants and
drawing up designs is easy,
but preparing the land itself,
whether the site is a woodland, a field or a portion of the
lawn, is usually backbreaking
work. Even the neophyte gardener knows that the existing
plants have to go—sod must be
skimmed, shrubs and trees up-
rooted—and some experienced
gardeners both clear the
ground and double-dig (at least
they say they do in books). I
don't have the time or the energy for all that work, and I've
found a way to avoid it but
still have a planted place.

When I moved to my cur-
rent home, there were two

acres of horse pasture, one
acre of lawn and not one cultivated planting. From my last
garden, I had brought
enough perennials to plant
half an acre. Rather than strip
and spade that much lawn, I
spread newspaper and mulch
on the grass and waited. To
my delight, I was able to
smother the turf and plant the
area to perennials, all in one
growing season. I've continued to convert my property
piecemeal to ornamental
plantings, each year tackling
one or more portions of the
horse pasture, the lawn and
the wilder parts along the
property line. Wherever I've
tried the combination of newspaper and mulch, it's worked.

I've learned that I can
smother almost any plant, including woody vines such as
poison ivy and bittersweet,
fleshy-rooted bank-holds such
as daylilies, and taprooted
renegades such as dandelions

and pokeweed. Most of the
work is light stuff, a few hours
of spreading newspaper and
mulch. I can smother hundreds
of square feet of plants at a
time with materials I have on
hand or can buy cheaply.
And I can plant right through
the newspaper and get new
plants started while the old
ones are dying. I've managed
to start bulb gardens this way
that looked full in less than a
year, and the technique has
worked just as well for starting
shrub borders, perennial borders and vegetable gardens.

Getting started

My first step is to lay out the
new garden, generally in late
fall or early spring. I push 5-ft.
bamboo stakes into the
ground wherever I think I
want a tree, and lay down a
garden hose to mark the garden's edge. Then I take time
to look at this design from van-
tage points around the prop-

erty and various windows of
the house, making changes
until I'm satisfied. I consider
both the shape of the garden
and its practicality—how to
maneuver the wheelbarrow
and lawn tractor around it.

When I'm sure about the design, I cut the existing plants to
the ground. If I can, I mow. If
the plants are too tall or too
woody, or are located in too
wet a spot, I scythe by hand. I
saw saplings off at ground level and then peel the bark on
the stumps to 4 in. beneath
ground level, which largely prevents suckers from sprouting. If a
sucker appears, I cut it off and
cover the stump with a 4-in.- to
6-in.-thick layer of newspaper.
On some sites, such as open
meadow and turf, nature has
already leveled the plants in
late fall or early spring. Then I
skip the mowing and trimming,
and go right to papering.

I spread whatever paper
products I have on hand. Usu-

ally I use newspaper because it's readily available and the pages are large. I've also used paper tablecloths, magazines, junk mail and recycled typing paper.

The paper layer must deprive the plants of sunlight completely. I overlap adjacent sheets at least 3 in. This keeps the smothered plants, which struggle for life, from rising to the sun between sheets. The paper layer must also be suitably thick. For turf, mowed fields and most weedy areas, I've been successful with a layer as thin as two sheets of newspaper, but I generally use four or five. The thicker the layer, the less likely it is to yield to sprouts, or to be torn by an errant child or pet. If I'm smothering superbly stalwart plants such as daylilies or honeysuckle, I use particularly thick paper layers, or cardboard.

When I moved here, I found an ancient thicket of wild orange daylilies interlaced with poison ivy for 60 ft. along a stone fence. Late one fall, when the plants were dormant, I cut the daylily flower stalks to the ground and then got out my saved copies of *The New York Times*. I covered every inch of the site at least three sections deep—a solid layer of roughly 30 sheets—and then spread a 2-ft.-deep layer of the neighbors' leaves on top, largely to hold the paper in place. The next spring, a total of six spindly daylily fans made it through the barrier. I killed them easily with a spot spray of herbicide. I saw no sign of the poison ivy. A year later, I dug down through the newspaper and found the remnants of the daylilies, dead-looking crowns with desiccated roots. I've left the newspaper in place and planted shrubs through it. Eventually, when the newspaper and leaves have decomposed, I'll plant a shallow-rooted ground cover among the shrubs.

If this had been just a turf site, I'd have used only two layers of paper. My lawn contains many different violets, a superb and free ground cover. Early on, I discovered that a two-sheet layer kills the grass but still allows the violets—and dandelions—to grow through.

Carney varies the newspaper layer, making it thickest—up to 30 sheets—for tenacious plants such as daylilies and poison ivy, and as thin as two sheets to smother turf, as shown here. Covered last fall, the dead grass at the center resists wetting. By next year, it will have broken down. If Carney plants at the stage shown here, she breaks up the turf at each planting hole.

Digging the dandelions later is easier and cheaper than setting a more expensive ground cover would be.

I also spread paper when I make paths. I lay out the path with a hose, then cut the plants to the ground. Next I cover the path with a ½-in.-thick layer of paper products, preferably cardboard. The thickness of this layer is important. I want no plants to survive. The thicker the cover, the longer it will take to decompose, and the longer before anything will grow again. For aesthetic reasons, I cover the cardboard with mulch.

I always cover the newspaper with an organic mulch. For one thing, the paper needs some sort of weight to stay put. (I try to work on windless days and cover a small section at a time— you never know when the wind will show up.) Also, the mulch eventually enriches the soil. I favor leaves because they break down within a season, cost nothing and look good. I've read that the leaves of some trees, such as maples, can lower the soil pH, but I've used sugar-maple leaves for seven years with no ill effects. I've also used sawdust, wood chips, hay, pine needles, scythed weeds, shredded cedar roofing shingles, grass clippings and the dog's cedar excelsior bed. If I use a mulch such as hay that I know to contain weed seeds, I cover it with something weed-free, usually leaves, as soon as I have enough. In my current project, I'm spreading maple leaves on top of hay. The leaves have maple seeds, and a few seedlings have popped up. I just pull them.

I try to match the mulch and the soil. If the soil pH is acid, I avoid pine needles and bark. On heavy clay soil, I've spread sand with the mulch; earthworms work it in eventually. My present soil is rich loam, so I use whatever mulch I have.

I leave the paper and mulch layers in place when I set out plants for a new garden. I usually plant so thickly that in a year or two the mulch is hard to see. If you don't like the looks of the layers, you can remove them to the compost pile once you're sure the previous plants are dead and few weeds are likely to germinate. Or, you can shred your mulch before you spread it, so it will decompose faster.

I vary both the mulch and its depth to suit conditions. On turf that children and pets rarely visit, I sometimes use just enough mulch to keep the paper down, even though some paper peeks through. On vigorous plants such as poison ivy, I pile the mulch at least 6 in. deep. I think anyone new to smothering plants is likely to mulch too lightly at first. If in doubt, err on the side of too deep. I make a very deep layer of leaves or straw—as much as 1 ft.—if I have an extensive supply. Heavy mulch is especially important in windy locations such as the Plains states or along a major body of water. Also, I find that plants push up more vigorously in spring than in fall, so I mulch deeper then. Moreover, in March we tend to get stronger winds that shift in direction, while in fall early snows help weight down the mulch and the paper.

I've seen paper and mulch improve poor soil in a few years, particularly when I renew the mulch annually. Seven years ago, I planted my current rose garden of 80 Hybrid Teas on a weedy, unused, stony driveway. Not only had I spread newspaper and covered it with wood chips, but the first fall I'd also spread leaves 2 ft. deep over the area. The following spring I dug bushel-size holes for the rose plants, and filled these holes with humus from my compost pile. I also planted hyacinths between the roses and have added more over the years. Every year, I've collared the roses in the fall, filled the collars with wood chips, and then piled 2 ft. of leaves over everything. I remove the collars in the spring and spread the wood chips over the rose garden to give it a neater look. Snow settles the leaves to just a few inches deep, and the hyacinths push through. The rose garden now has a thick rich topping of mulch, worm castings and humus. Some of the hyacinths are seven years old and still blooming magnificently. I could have spaded up the driveway, but I doubt the results would have been as satisfactory. If I had rototilled, weeds would have been a misery for several seasons. And if I had double-dug, my back would still be aching.

Mulch alone is not enough to hold down the newspaper. Along the edge of a new garden, I place weights atop the mulch at 2-ft. intervals to prevent the wind from gaining a toehold. If the garden is more than 4 ft. wide, I also lay weights at 2-ft. to 3-ft. intervals across it. I use logs from my woodpile and branches that have fallen from the trees. They'll decompose eventually, so I don't remove them from the garden. When I start to plant, I may move them around to act as markers between my new perennials and shrubs. Once the perennials emerge, the markers are invisible. On other sites, I've advised gardeners to use rocks, bricks, plastic containers filled with water, and toys.

Edging

Lately, when making new gardens from turf, I've been edging before I lay paper and mulch. Edging at the start saves work later, helps whoever is mowing the adjacent lawn and makes the garden's shape clearer. I lay out the garden hose as usual, but then I slice the turf alongside it with an edging spade and cut a second line parallel to the first, 3 in. to 6 in. away. I remove the sod between the lines—in the spring it rolls right up. I often use the sod to fill in low spots in the yard. Otherwise, I toss it on the compost pile, or place it on top of sod I intend to paper, and smother it.

Six months before this photo was taken, this expanse was a lawn. Carney planted daffodils and other bulbs into the grass through paper and mulch hay.

Pulling the sod leaves a trench in the lawn that's more visible at a distance than the garden hose is. It gives me another chance to review the shape of the new garden, which I think is time well spent. I once laid out a rhododendron garden to separate the front and rear yards, with a 3-ft.-wide pathway through it. The front yard has stone fences or stone steps on all sides. My husband approved the design until he saw the trenches and realized that his 48-in. tractor couldn't get to the front yard.

I lay the paper to the middle of the trench. The result is a clean line that shows the grass cutter where to cut, so wheels and blades are less likely to tear up my work. The edge also looks neater because no uncut grass grows rankly along it. With the first rain, the paper bends into the trench and stays there, mak-

ing it harder for winds to get under the edge.

I sometimes make the trench deeper with the spade or a small pickax, and set permanent edging between the new garden and the turf. This is heavy work but worth it. It spares me the annual chore of cutting back the lawn where it creeps into the garden, and it keeps the lawn mower from straying, too. I like to use 6-in. black-plastic edging with a rolled top. It looks good and lasts forever.

Planting

When I paper and mulch a new garden in the fall, it's ready to plant the next spring. Any other time of year, I wait at least three months. To set a plant, I part the mulch, cut through the paper and dig a planting hole. Sometimes I find smothered turf that has the consistency of dry peat moss. I've found that it resists

wetting and will deprive a newly set-out plant of moisture. To avoid problems, I turn over and chop up at least one spadeful of the turf and set the plant in the middle of the disturbed soil. By the second year, the turf has broken down and no longer resists wetting, so I can add plants without as much digging. Once I've made a satisfactory planting hole and set out the new plant, I pull the mulch back around it. Weeds rarely come up to compete.

I sometimes plant bulbs in the fall directly into turf, and then spread a light layer of newspaper and mulch, enough to smother the grass but too little to hold back the bulbs. I was inspired to try this method when I failed to smother an onion and chive bed one fall with three sheets of newspaper and mulch. The onions and chives pushed right up the next spring. It seemed to me that other bulbs might behave the same way, so I planted auratum lilies in a patch of grass and then covered the patch with three sheets of newspaper and a layer of mulch. The next spring, the lilies came up and the grass did not. That fall, I laid out another garden on turf and planted a few hundred daffodils, smaller spring bulbs like spring glories and snowdrops, and delicate spring bloomers like bloodroot, trilliums and Virginia bluebells. Then I spread paper and piled about a foot of maple leaves over it all. Everything I planted came up the next spring, and the effect resembled a well-established garden.

If you're eyeing a neglected weedy portion of your property, or a piece of your lawn, and dreaming about a new bit of garden, I encourage you to try starting with paper and mulch. You'll put waste materials to use for little or no money, save yourself a lot of labor and get a start on improving your soil. Best of all, you can go straight to the fun—selecting and setting out new plants. □

Nancy Carney lives in Newtown, Connecticut, amid a shrinking lawn and growing garden.

Peat Moss
Does this widely available soil amendment have uses in your garden?

Peat moss joins perlite, topsoil and sand in a homemade potting mix. Super absorbent and long-lasting in most garden soils, peat moss has long been valued by gardeners as an organic soil amendment and as an ingredient in potting mixes.

by Warren Schultz

When I was younger, I begrudged peat moss its popularity. The day I dreaded most while working at our family greenhouse and nursery was the one in early spring when a tractor trailer, filled front to back and top to bottom with peat moss in wet, frost-slicked, 150-lb., 6-cu.-ft. bales, whooshed to a stop just under my bedroom window. The day didn't end until that trailer was empty and our footsteps echoed inside of it. After only an hour or two of toting, I'd find myself working to the muttered mantra of "What do people want with all this &*#$ peat?!"

Despite my early experiences, I didn't swear off peat moss. In my 30-odd years of gardening, I have often found occasion to use it as a soil amendment and as an ingredient in potting mixes. I still use it today, though more sparingly. I'm aware now of some weighty environmental issues surrounding the harvesting of peat moss, but for some applications, it's hard to beat peat.

Photo: Susan Kahn

What is peat moss?

Peat is organic matter that has accumulated beneath the surface of poorly drained areas called bogs. Because the organic matter is constantly saturated with water, normal decomposition (which requires oxygen) can't take place. The organic matter does decompose, but only very slowly, over the course of hundreds or even thousands of years.

Peat can be made of the remains of a variety of plants, but by far the most common constituents of peat are sphagnum mosses—small, primitive plants that thrive in the spongy, acidic growing conditions typical of bogs. As moss plants die, they fall below the surface of the water, where they slowly decompose into peat moss. To harvest or mine peat moss, peat companies bring in a succession of machines to strip off the spongy upper layer of live moss and slightly decomposed peat, and then harrow, dry, slice, vacuum and package the peat moss.

There are at least four distinct products harvested from a peat bog. Live sphagnum may be sold green or dried as a lining for hanging baskets or as a potting medium for orchids, which thrive in a loose, well-drained mix. Beneath the live moss is peat in various stages of decomposition. The light, long-fibered peat moss found near the top of the bog is the stuff you find in plastic-wrapped bales at nurseries. The coarser material from the middle layer is often called poultry peat because it's used as chicken litter. Finally, from the bottom, there is peat moss that has decomposed almost completely into a dark, crumbly, soil-like substance sold as peat humus, which gardeners sometimes use as an alternative to bagged topsoil.

The advantages of peat moss

Twenty years ago, peat moss was the soil amendment of choice. In recent years, other organic materials such as shredded bark and compost have challenged its supremacy, but peat moss still has things going for it that other soil amendments don't.

Peat moss is super-absorbent. The cells of sphagnum moss are large, rectangular and nearly empty, so they can absorb and hold up to 20 times their weight in water. Peat moss mixed into light soils therefore helps the soil retain moisture longer.

Peat moss is relatively stable. Its cell walls are made of rigid materials that are slow to decay, so peat moss

retains its structural integrity for a long time, even while wet (sort of like those cereals that stay crispy in milk). While manure can break down in weeks and compost in months, peat moss can take several years to decompose in garden soil. (Shredded bark is even slower to break down, making it the amendment of choice in the warm, moist soils of the South, where even peat moss breaks down rapidly.) As peat moss decomposes, it continues to improve the tilth, or friability, of the soil, making more space for air and water to penetrate.

Unlike manure and compost, peat moss is sterile and weed-free. Weed

A clump of moss pulled from the top of a bog in eastern Canada reveals a tangle of tiny sphagnum plants. As old plants die, they slip below the surface of the watery bog, decomposing slowly over hundreds of years to form peat moss.

seeds and disease-causing organisms can't survive in a waterlogged bog, so you don't have to worry about importing unwelcome guests when you buy a bale of peat moss.

Perhaps peat moss's biggest selling point is the characteristic that imprinted itself on me in my formative years. Peat moss is portable. It's packaged. It's everywhere. Virtually every garden center and hardware store in America starts the spring with a mountain of brightly packaged peat moss. In addition to the large, heavy bales, peat moss is often found these days in tiny bales—as small as ½ cu. ft.—with handles.

Peat moss in the garden

Transplanting trees and shrubs—Back in the days when I was toting all those big bales, most of the peat moss was probably being used to amend the soil in planting holes. In recent years, however, scientists have questioned the value of routinely amending soil before planting trees and shrubs. The latest research suggests that amendments can be a good thing, but only under certain conditions: the plant prefers soil rich in humus (azaleas and rhododendrons are examples); the hole is made big enough—at least five times the diameter of the root ball; the amount of organic matter added is not excessive—no more than one third of the backfill; and some amendments are also worked into the surrounding soil.

For best results, backfill a planting hole with a mixture of 1 part peat moss to 2 parts soil or spread a 2-in. deep layer of peat moss over the planting area and rototill it in to a depth of 8 in. before planting.

Amending soil in flower beds and vegetable gardens—In a perfect world, I'd always have enough of my own rich, homemade compost to be able to add a thick layer of it to every new garden bed. I don't. The next best thing is peat moss. So, when I prepare a new bed, either for flowers or for vegetables, I apply a 2-in. layer of peat moss and till it in to a depth of 6 in. to 8 in. The peat moss increases the moisture-holding capacity of the soil while at the same time improving soil structure and drainage.

Making potting mixes—Here's where I'll accept no substitute for peat moss. Because it's sterile, lightweight, absorbent and fast-draining, it's an ideal ingredient in potting-soil mixes, especially those used for starting seeds. I make my own seed-starting medium by mixing 1 part each of peat moss and vermiculite. Then I add 1 cup of a granulated natural fertilizer per bushel of mix.

The disadvantages of peat moss

Peat moss is not perfect. It carries virtually no nutrients—less than 1% nitrogen and even less phosphorus and potassium. If a soil test reveals that your soil is deficient in key nutrients, you'll have to sprinkle fertilizer over the peat moss before you turn it in.

Peat moss is acidic. It has a pH of from 3.5 to 5.0, so adding a generous layer can acidify your soil by a point or more. Unless you have alkaline soil or you want to lower the pH for acid-loving plants such as blueberries or azaleas, you need to add about 5 lbs. of lime per cubic yard of peat moss to counteract its effect on soil pH.

The convenience of peat moss comes at a price. Here in northwestern Vermont, a 2-cu.-ft. bale costs anywhere from $6 to $8. That's not bad compared to potting soil or seed-starting mix, but it's certainly more expensive than homemade compost or even store-bought manure. And it's one and a half to two times more expensive than shredded bark or manure bought in bulk.

Finally, there's the environmental cost. We can all work up a warm feeling about using compost or manure, because we're putting waste to work in the garden. But peat moss is a natural resource that is renewable only over hundreds or thousands of years. Harvesting peat moss also disrupts wetlands, displaces wildlife and destroys other native plants. In addition, peat bogs are carbon sinks. As bogs are mined, they release carbon dioxide, contributing perhaps to global warming.

The Canadian Sphagnum Peat Moss Association, an industry group, maintains that peat moss is a renewable resource. The association encourages peat moss companies to preserve native bog plants and to reclaim and restore bogs once they've been harvested. The association also says that the industry is barely scratching the surface of peat moss reserves. Canada, the source of virtually all the peat moss used in North America, has 274 million acres of peatland, but less than 0.02% of that area is under harvest for horticultural peat moss. In fact, new peat moss is forming faster than it's being removed; 50 million metric tons of peat accumulate annually, while fewer than one million metric tons are harvested within the same span of time.

So, it seems that peat moss presents a dilemma of our modern times. We as gardeners have to weigh the environmental consequences against our desire to improve a patch of soil or give a tree or shrub a better chance of surviving and maturing in our backyards. □

Warren Schultz is former editor of National Gardening *magazine. He gardens in Essex Junction, Vermont.*

Visitors to a bog in Newfoundland, Canada, keep their feet dry on a winding boardwalk. Though inhospitable to trees, a bog supports an amazingly diverse community of plants.

A peat bog under harvest resembles a newly plowed field. Peat companies strip away the living vegetation, dig ditches to drain the bog, then repeatedly harrow and vacuum the uppermost layer of peat moss as it dries in the sun.

Ready for shipping, a mountain of plastic-wrapped, 4-cu.-ft. bales of peat moss fills a warehouse. Peat moss is widely available and readily portable, making it more convenient to purchase and transport than soil amendments bought in bulk.

Photos: top, Val Wilkinson/VALAN; center and bottom, Chris Curless

No-Till Gardening

Soil improvement from the top down

by Lee Reich

I grow most of my vegetables and fruits by a method that minimizes weeding and creates an ideal home for plant roots. I've laid out my garden in permanent beds, and every year I mulch them with compost, which suppresses weeds, retains moisture and also nourishes my plants. Under the mulch, a healthy population of creatures "plows" the soil and continually improves its structure. I adopted this system nearly 20 years ago, and have used it since then on a variety of soils, including clay loam in Wisconsin, sand in Delaware, rocky clay in Maryland and my present silt-loam in New York.

My technique saw its beginnings when I was in graduate school studying soil science. I took in the standard classroom fare, but I also had the good luck to come across three books in the university library that appealed to the iconoclast in me: Each, in its own way, questioned the sense of tilling the soil. In *Plowman's Folly* (published in 1943; reissued in 1988, $21.95 postpaid from Island Press, P.O. Box 7, Covelo, CA 95428), Edward Faulkner suggested that farmers, instead of plowing the land each year, compare their fields with nearby fencerows, where the soil, protected by plants and never plowed, was invariably of superior tilth. Buried weeds remain dormant as long as they're not exposed to light and air. Why turn over the soil and bring buried weed seeds to the surface? he asked. He advocated leaving the soil's natural layers untouched and building fertility through the use of cover crops knocked down and left on the soil surface. Ruth Stout, in her wry classic *How to Have a Green Thumb without an Aching Back* (published in 1956 and available at many bookstores), gave a prescription for nearly effortless vegetable gardening: Pile the soil deep with a year-round mulch of straw or hay. She teased the mulch apart to sow seeds, settled it back around the young plants, and sailed through the season without a crop of weeds to pull. Also in 1956, Rosa Dalziel O'Brien published *Intensive Vegetable*

Author Reich grows vegetables and fruits with a system of year-round mulch, permanent beds and as little disturbance of the soil as possible. The benefits are good soil structure and fertility, and nearly weed-free gardening. In the seven-year-old bed above, Reich can pull up the entire taproot of a young dandelion without digging. For mulch, he spreads 2 in. of compost on his garden beds annually. The top 3 in. of soil is rich in organic matter, and feeder roots proliferate in this fertile zone (below).

Culture (now out of print), a description of growing commercial vegetables in England on soil that was never tilled, but was covered periodically with a blanket of compost.

Tradition dies hard. The spring after I'd read those books, I set out to convert a patch of lawn into a vegetable garden, with the help of two shovels and a friend with a strong back. We inverted each shovelful of soil, burying the sod to kill it. Then I chopped the large clods with a hoe, raked the soil smooth and sowed seeds.

A frightening number of weeds popped up on the heels of the vegetable sprouts, just as Faulkner had predicted. What to do? Hoeing or rototilling would damage the shallow roots of my vegetable plants. By luck, I ran into a city maintenance crew that had just cleaned water plants from a nearby lake. They dumped a truckload of the plants in my driveway. I laid pitchforkfuls between the rows of vegetables and managed to smother the young weeds. I decided then that it was time to abandon tradition and follow the logic of Faulkner, Stout and O'Brien.

Three principles

Avoid compacting—The first principle of my method is to stay off the soil to avoid compacting it. I never tread on my permanent beds, and I'm fussy about visitors sticking to the paths. Most of the beds are 3½ ft. wide, a size that allows me to stay on the paths and still comfortably reach the center of the beds. I make narrower beds along fences so I can care for them from one side.

Mulch—My second principle is to mulch the beds with compost every year. I usually spread the compost in fall, so it can meld a bit with the soil before spring planting, but some years I've done it in spring or summer. I generally apply 2 in. of compost each year, but I'll put down more if I have it. From frost to frost, my growing season is about 150 days, and summer temperatures are warm but rarely hot. In gardens with a longer, hotter growing season, where compost would break down faster, you might need several inches more a year than I do.

Two inches of compost a year should supply my plants with all the nutrients they need, but for insurance I've occasionally added a bit of nitrogen, phosphorous and potassium in the form of soybean meal (7% nitrogen and available at feed stores), rock phosphate (2% available phosphorous) and wood ashes (7% potassium). My present garden is seven years old, and the soil, which was fairly good when I started, has improved dramatically. If I were starting fresh on poorer soil, I'd probably add fertilizer the first year or two to be sure of good crops, though I'm not sure it's necessary.

I make literally tons of compost each year. A cubic yard, which weighs roughly 1,000 lb., covers 150 sq. ft. of soil when spread 2 in. thick. My main vegetable garden totals 600 sq. ft., and I also apply compost to my fruit trees, as well as leaf mold and grass clippings. I normally have four or five compost piles in progress, and as I use one I start another. I assemble a 4-ft.-square crib of planks notched and stacked log-cabin fashion, and fill the crib 5 ft. high with garden wastes, vegetable scraps from the kitchen, sprinklings of soil and soybean meal, large quantities of hay, and horse manure, which I haul free from a local stable. Then I dismantle the crib, reassemble it nearby and build another pile. In a month, I reassemble the crib and turn the two piles into it. If I lacked hay and horse manure, I would shred and compost leaves, or simply mulch the beds with an inexpensive, easily spread material such as sawdust and add an annual dressing of fertilizer.

To keep weeds out of the paths between the beds, I mulch deeply with just about anything (except compost). Coarse, slow-to-degrade materials are ideal, since they need less frequent replenishment and provide good footing even when the weather is soggy. I've used layers of newspaper topped by wood chips, sawdust, and old pieces of plywood.

Minimize disturbance—The third principle of my system is to minimize disturbance of the soil to avoid unearthing weed seeds and to preserve the soil's tilth. I don't till or dig the beds, and I use care when weeding, harvesting and cleaning up the garden in the fall.

I pull most weeds by hand, and they come up roots and all. For the occasional dense stand of shallow-rooted weeds, I grab the tops with one hand and cut the roots just below the soil surface with a Cape Cod weeder, which has a blade at a right angle to the handle. For taprooted weeds, such as dandelions and docks, I slide a shovel alongside the root, and rock the shovel as I lift the plant. I can pull up foot-long roots without leaving a portion in the soil to regrow, and without turning over the soil. On Ms. O'Brien's advice, I let all the weeds grow a bit before I yank them. They're easier to handle, and I get fodder for the compost pile.

I also minimize soil disturbance when harvesting. I slice off young lettuce and spinach plants with a knife at soil level. I give older plants a quick twist to sever the small roots, and then pull the plants out of the ground. If the soil is moist, I can pull up carrots by their leaves. In dry soil, I slip a shovel or trowel alongside them, as I do for stubborn dandelions. I use the same method for parsnips, which produce much bigger roots than carrots do.

When it's time to pull the old, woody stalks of plants such as broccoli and corn, I first snip the top off each plant, leaving a foot-high stump for a handle. Then I cut the lateral roots by slicing straight down into the soil around the base of the plant. Finally, I twist the stump and it comes free.

Starting a no-till bed
I used to start new plots by mowing the existing plants close to the ground, strip-

Reich makes several compost piles yearly of garden wastes, vegetable scraps from the kitchen, hay and horse manure. He'll disassemble the crib of notched planks shown here once the pile has settled and will hold its shape, and then reassemble the crib to start a new pile.

ping them off with the top inch of soil, spreading compost, and planting. A few years ago, I decided to leave the soil alone, provided the drainage was adequate. Now I just mow, spread newspaper to smother the existing plants, cover the newspaper with compost or a layer of sawdust, and plant. The soil structure remains intact and the smothered plants decompose to enrich the soil. (For more on this method of breaking ground, see "Mulch, Don't Dig" on pp. 68-70.) It's easier to begin the first season with plants, rather than seeds, but by the second season, starting with seeds gives good results.

There are times when I dig or till. For example, when I'm transplanting trees, I dig a hole just big enough to accommodate the roots, then spread newspaper and mulch to smother existing plants out to the tree's dripline. In another case, if a soil is so poor that even weeds struggle to grow on it, I either look for a better place to start a new garden, or amend the soil. A few years ago, I broke ground in a vein of heavy clay soil. Weeds were doing fine there, but the drainage was poor. To improve the drainage, I covered the site with a 6-in. layer of sawdust and dug it in thoroughly. (Peat moss would also work.) Since fresh sawdust can draw nitrogen from the soil as it decomposes, and acidify the soil, I mixed 2 lb. of soybean meal and 1 lb. of limestone with every 5 gal. of sawdust. I haven't dug the soil since. The sawdust has loosened the soil nicely, and by the time it decomposes, years from now, there will be a permanent improvement in soil structure and plenty of new organic material brought down from the surface by earthworms.

I dig or till on one other occasion—when I grow crops that require a lot of digging, such as potatoes and sweet potatoes. Since I have to disturb the soil at harvest anyway, I treat these crops more conventionally: I till their beds once, adding manure and compost; plant the crop; and then hoe lightly the rest of the season or mulch heavily to suppress weeds.

Taking stock
It may be satisfying to look back over freshly plowed, rototilled or spaded soil, but in my view leaving the soil intact brings me greater benefits. I can plant cool-weather crops like peas earlier in the spring than my tilling neighbors can because I don't have to wait for the soil to become dry enough to work. My plants get the most from rainfall because it easily penetrates the soil surface, and because leaving the soil undisturbed preserves capillary connections that wick water to the roots. As the compost mulch decomposes, it enriches the top few inches of soil, the ideal zone for plant nutrients, since, to quote Faulkner, "...it is the tiny, tender feeding roots in the surface layers of the soil that do the real work of finding food." Thanks to the mulch, those roots also rarely suffer from heat or moisture stress, even during summer droughts. My soil escapes two effects of digging and tilling: excessive aeration, which accelerates the breakdown of organic matter, and disruption of the soil structure. Finally, I'm not a slave to weeding. In fact, at times I wander out into the garden, bucket in one hand and Cape Cod weeder in the other, only to be disappointed at the dearth of weeds. □

Lee Reich gardens in New Paltz, New York.

Gardening amid Tree Roots

Understand the risks before you dig, water or apply herbicides

Painstaking digging and a can of whitewash reveal a network of tree roots only inches below the forest floor. The tiny feeder roots at left grow up into the porous topsoil from the larger transport roots at right. Contrary to expectations, tree roots in most soils grow just below the surface.

by Thomas O. Perry

I knew I had finally become an expert when my dad asked me for advice. The foliage on his pet dogwood tree was wilting, and several branch tips were obviously dead. He asked anxiously, "Does the tree have borers? Should I send a leaf sample to the plant pathology lab at the university?" I shook my head sadly and responded, "No, Dad. You injured your dogwood when you tore up its roots to plant 144 daffodil bulbs."

Like many gardeners, my father failed to understand that gardening affects the health of nearby trees. Tree roots are closer to the soil surface, wider-ranging and more vulnerable to damage than gardeners realize. Most tree roots are concentrated in the top few inches of soil where they compete directly with the lawn and garden plants for moisture and nutrients. When you dig, water, apply herbicides or compact the soil around trees, you can injure or even kill them.

I'm not suggesting that you stop gardening, but I do hope you'll

Photos: Thomas O. Perry

consider your trees when you garden. The following basic information and rough guidelines, which come from my years of work as a tree researcher, will help you to have both an attractive garden and healthy trees.

Some facts about tree roots

Most tree roots are small (thinner than the lead in a pencil and less than 2 in. long) and are concentrated in the top few inches of the soil. They are non-woody and ephemeral, growing and dying the same year. Often called "feeder" roots, they absorb most of the water and minerals a tree takes up from the soil. They grow outward and upward from a maze of woody, perennial transport roots, which vary from ¼ in. to 1 in. in diameter. Transport roots usually grow parallel to the ground; most are 4 in. to 12 in. below the soil surface. In turn, they branch from large, shallow roots that radiate from the base of the trunk. There are usually four to 11 of these large roots, which may be several inches in diameter and nearly as old as the trunk.

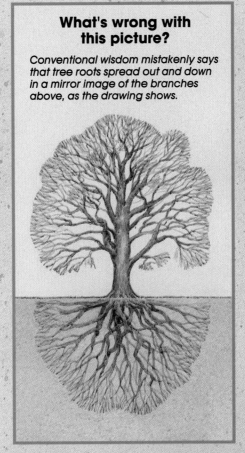

What's wrong with this picture?

Conventional wisdom mistakenly says that tree roots spread out and down in a mirror image of the branches above, as the drawing shows.

Tree roots, like all roots, are opportunistic. They grow only where competition with other roots is not too fierce and where the requirements of life—moisture, nutrients and, above all, oxygen—are readily available. Because these life-sustaining elements tend to be most plentiful in the uppermost layers of the soil, commonly called topsoil, most tree roots grow near the soil's surface. In typical clay-loam soils, 90% of a tree's roots are concentrated in the upper 18 in. of soil. If conditions permit, tree roots will grow downward, occasionally penetrating to great depths, but in most soils the primary direction of root growth is horizontal and up into the porous top soil, *not* down into the compacted and often water-logged subsoil.

Tree roots cover more ground than you might think. The rope-like transport roots normally extend far beyond the branch tips of the tree and weave over and under the roots of other plants as they travel horizontally through the soil. Indeed, it's not uncommon for the root system of a tree to occupy an area several times larger than the

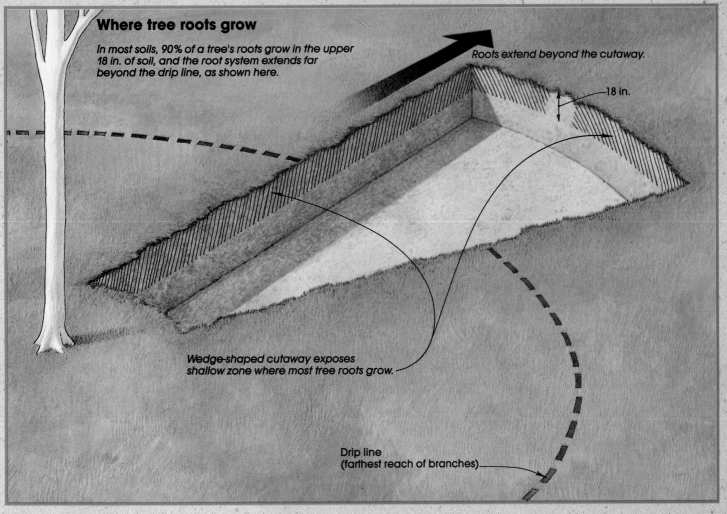

Where tree roots grow

In most soils, 90% of a tree's roots grow in the upper 18 in. of soil, and the root system extends far beyond the drip line, as shown here.

Roots extend beyond the cutaway.

18 in.

Wedge-shaped cutaway exposes shallow zone where most tree roots grow.

Drip line
(farthest reach of branches)

Illustrations: Gary Williamson

circle defined by drawing an imaginary line from the branch tips to the ground (the familiar "drip line"). The roots of mature trees growing in open areas typically grow 40 ft. or more beyond the drip line. My champion tree root extended along a crack in the pavement for 117 ft. from the tip of the tree's nearest limb!

You may be surprised that a tree's root system is a shallow network extending far beyond the drip line. After all, conventional wisdom says that roots mirror the canopy, growing deep and wide in a pattern like that of the branches above them. But tree roots simply don't forage for water and nutrients many feet below the soil's surface, as we've been led to believe. Even those trees, such as oaks, pines and walnuts, that begin life with a pronounced tap root, have a dominant horizontal root system running just below the soil surface.

All parts of a tree are interdependent. Leaves can't photosynthesize essential sugars without vital supplies of nutrients from the roots, and roots can't function without sugars and other supplies from the leaves. If some of the leaves die, then some of the roots will die. If some of the roots die, then some of the leaves will die. At equilibrium, a typical tree is (by weight) about 5% leaves, 5% absorbing roots, 15% twigs and branches, 15% transport roots, and 60% trunk. Gardening and other activities that disturb this equilibrium will lead to a decline in tree health.

When tree roots are damaged, a corresponding number of branches will ultimately die, rot and shower down on the house, on the car in the driveway or on the heads of the gardener and his loved ones. If the damage is too great, or if it's repeated year after year, the entire tree will eventually die. Some branches will die during the first year or so after the damage is done, but the decline is often gradual, and the tree may take ten years or more to succumb.

If you're dismayed by now because you have a tree whose roots have been damaged recently, take heart. It's possible to reverse the process by mulching around the base of the tree, watering for two or more growing seasons, adjusting the soil pH to the taste of the tree (most trees prefer a slightly acid soil) and modifying the garden

Roots killed by trenching

Trenching near a tree can kill from 10% (top) to nearly half (bottom) of a tree's roots.

Root spread

Drip line
(farthest reach
of branches)

Trench

Trunk

Roots
killed

plan to allow tree roots to remain undisturbed. However, dead branches can't be brought back to life, and recovery is usually slow. Several of the trees I've worked to reinvigorate have needed as many as three years to develop healthy foliage and more than eight years to grow a new set of branches.

Gardening around trees

Now that you know how and where tree roots grow, you can understand why many gardening activities can harm trees. Here are some basic guidelines to help you garden without causing excessive damage to your trees.

Digging—When you consider putting in a new flower bed, think before you dig. Site the bed out of reach of tree roots or at least far enough from the tree that your digging will do relatively little damage. The closer to the trunk you dig, the more roots you separate from the tree. When you prepare the soil for a flower bed under the canopy, you are depriving the tree of a significant portion of its moisture- and nutrient-gathering ability. Extensive branch dieback will soon follow. In an extreme case, you can kill the entire tree, as my father did. I plant daffodil and crocus bulbs under my trees, and there is nothing wrong with putting in a few small flowering trees or shrubs. The key is to plant only a few at a time. Trees can tolerate modest plantings each year, but you can't tear up the root systems of trees repeatedly and promiscuously to establish formal gardens beneath them.

If digging on a small scale can do great harm to tree roots, imagine the threat posed by large-scale landscaping and construction. In a matter of minutes, a backhoe can cut away huge sections of a tree's root system. Extensive grading can suffocate tree roots by raising the soil level, or it can tear away the equivalent of miles of tiny absorbing roots by lowering the soil level. The installation or repair of underground utilities running near the base of a tree requires trenching that may cut off major roots and the thousands of transport and feeder roots that are connected to them.

Consider the alternatives before unleashing a bulldozer or a trench digger. The costs of removing dead limbs from a favorite oak are usually much greater than the savings gained by taking the shortest distance between two points.

Soil compaction—Plants need porous soil. They don't have hearts to pump oxygen to their roots. Instead they rely on the millions of holes created by nature's plowmen: worms, ants, millipedes, moles and other creatures. Cracks generated by multiple cycles of freezing and thawing or of drying and wetting of the soil also serve as passageways for oxygen. Soil compaction, on the other hand, which can result from the traffic of children or pets or gardeners and their carts, closes the pore spaces in the soil, making it difficult for the roots to penetrate. The effect is to suffocate tree roots and reduce the area in which they can grow.

The most effective technique for minimizing the damage done by soil compaction is to limit intensive activities to areas of the yard where there are no tree roots. You can relieve compaction by leaving leaf litter in place as an energy supply for the creatures of the soil or by having a husky teenager drive the tines of a spading fork into the ground every 8 in. or so and rock the handle back and forth gently to fracture the soil around your trees.

Watering—It's important to water your trees like any other plant in your garden, but it's just as important not to overwater them. Contrary to the old saw, watering deeply will not cause roots to go deep into normal soils. Overwatering forces oxygen from the soil, causing tree roots and the creatures of the soil to suffer. Only plant pathogens thrive under such conditions.

It's better to water lightly and more frequently than to overwater. A typical good garden soil can hold about 1 in. of water in the top 10 in. of soil. On a hot summer day, as much as ¼ in. of moisture can evaporate away; during a hot spell, the reservoir of available water can disappear in just four days. Under such conditions, I recommend that you irrigate your trees and other garden plants with ¼ in. of water twice a week. Use pie plates set within reach of your sprinklers to measure the quantity of water your are applying.

To determine whether you need to water, pinch the surface inch of soil in the yard just as you would pinch the soil of a potted houseplant. If it feels dry, water. If it feels wet, don't water. When in doubt, *don't* water.

Raking and mulching—Raking and removing fallen leaves and needles to keep lawns and gardens from being smothered is harmful to trees. In the forest, fallen leaves remain on the ground and serve a variety of useful purposes. Leaf litter acts as a blanket to reduce competition from other plants, to keep the soil from freezing (which causes feeder roots to die back) and to slow the rate of evaporation of soil moisture. Leaf litter also supplies energy to soil aerators such as worms, and returns to the soil nutrients essential for plant life.

If you feel you must rake, then replace some of what you remove by mulching and fertilizing your trees. Apply mulch to a thickness of no more than 2 in. to 3 in. (taking care not to

Tree roots require soil loose enough to offer a good supply of oxygen. Confronted with a compacted soil smothered by bricks, the tree roots in the photograph above had nowhere to go but into the crevices between the bricks.

mound the mulch up against the trunk) inside a circle at least 8 ft. to 10 ft. in diameter. Smaller circles have the salutary effect of keeping lawn mowers away from the trunk, but they do little to replace leaf litter. Test your soil to determine the nutrient balance and the amount of fertilizer required, then broadcast the fertilizer from the trunk to at least 20 ft. beyond the drip line. Since tree roots grow in the surface layers of the soil, right along with the roots of the lawn and garden plants, there's no need to place fertilizers in holes bored into the ground. (See "Fertilizing Trees Makes a Difference" on pp. 58-60.)

Applying herbicides—Herbicides can kill trees as well as the weeds they are intended for. Herbicides such as those found in "weed-and-feed" mixtures, which are designed to kill

broad-leaved weeds, can also injure or kill nearby trees. From an herbicide's point of view, trees are "broad-leaved weeds" when they grow in a lawn. Follow the label instructions to the letter, and steer clear of the root zones (not just the drip lines) of your trees. I use herbicides routinely in my yard, but I apply them only by wiping them on target plants. I never apply them in a way that would allow these useful but dangerous chemicals to penetrate the leaf litter or come into contact with the surface of the soil.

Be careful with herbicides, and ask your neighbors to be careful, too, because tree roots often extend from one suburban lawn to another. I've seen all of the leaves fall off yellow poplar trees within ten days after the lawn next door had been treated with herbicides applied to kill clover. You should also be wary of spraying weeds in the cracks of a path or sidewalk if trees are nearby. The photo at left illustrates why.

Compromise gardening
How much damage from digging, herbicides or competition can an individual tree withstand? Nobody can say for sure. The quantities of light, minerals and water required for optimum growth are not precise. Also, plants have evolved the ability to sustain limited damage and recover by generating new roots and branches to replace those that have been lost.

But the extremes are easy to define. You know you're pushing the limits when you rototill a tree's root zone, raise the pH higher than is normal for the region, use herbicides to eliminate all broad-leaved weeds in a lawn or dig up every square foot of an area to eliminate tree roots. You have to compromise, but what you decide to do or not do depends on your personal preferences. Knowing where the roots of your trees are and what they need will allow you to do well by your trees and your garden. ∎

Thomas O. Perry taught in the School of Forest Resources at North Carolina State University for many years. He now owns Natural Systems Associates, a consulting firm in Raleigh, North Carolina.

Managing Water in Arid Gardens

Simple, efficient, economical methods

by Daniela Soleri and
David A. Cleveland

In some ways, we've got it easy gardening here in Tucson, Arizona. There's plenty of sun, the winters are mild, and we encounter very few weeds, insect pests or diseases. We garden throughout the year, although we plant different crops at different seasons (see sidebar, p. 83). By far our major concern is watering. Tucson lies in the Sonoran Desert. Our average annual rainfall is 11 in., with half falling during July, August and September. We usually get less than 5 in. from October through March, and less than 1 in. from April through June—our driest months.

To grow vegetables at all, we have to water, but that's not the easy solution people once thought it was. Tucson is the largest North American city that relies solely on groundwater for its municipal water supply. Water rates here are increasing, while our water table is dropping rapidly, so water conservation is a big issue for gardeners in this area.

During the hot season, gardeners may find their plants wilting in the afternoon after a morning watering. This is especially discouraging for newcomers who've recently moved to the desert from cooler, moister climates. The prospect of watering twice a day, combined with poor yields due to heat and other water stress, discourages many, who conclude that gardening in Tucson is not worth the money or the effort. Others persevere, only to be astounded by the amount of water they use.

It doesn't have to be that way. During the past 15 years, we've developed a way of gardening that uses water efficiently (returning about $8 worth of vegetables for each $1 spent on water), provides a steady supply of fresh vegetables and fruits, and looks like a peaceful oasis. We're both anthropologists, and our special interest is small-scale food production in dry areas, so many of our ideas come from our travels to study gardening methods in arid West Africa, Egypt and Mexico.

Only a minor fraction of the water supplied by rainfall or irrigation is absorbed by and stored in plants. Most of the water runs off the surface, percolates below the plants' roots, evaporates from the soil or evaporates from the plants' leaves. Our goal is to minimize the water that is wasted or lost from a garden, while supplying just enough for the plants to grow well.

Sunken beds—The most noticeable feature in our garden is the sunken beds. Each bed is surrounded by a berm, which serves as a walkway; the soil inside is at or slightly below ground level. We plant everything we grow—annual vegetables and flowers, trees, and perennials—

The authors grow vegetables, flowers and fruit year round in Tucson in a garden designed for efficient water use. The adobe-like walls of their sunken beds (above) channel rainfall or irrigation water into the planting area inside the bed. Windbreaks (like the one on the facing page) also help cut water loss.

The sunken beds are excavated to a depth of 18 in. to 24 in. to remove the heavy clay soil and the underlying caliche. Then they're refilled with a mixture of topsoil, compost and composted manure.

Soleri uses a hose with a bubbler attachment to flood the beds about 3 in. deep. This takes roughly five minutes, and supplies enough water to thoroughly wet the soil in the bed.

in these beds. The effect is a beautiful waffle pattern (covering about 600 sq. ft.) of earthen walkways transecting lush patches of garden crops.

Though strange to those accustomed to raised-bed or row gardening, these sunken or basin beds make sense for water conservation. In desert climates, the rate of water loss through direct evaporation from the soil to the air is very high. The drier and hotter the air, the more water vapor the air is able to hold and the greater the evaporation from garden soil. Wind increases the rate of evaporation by blowing away humid air near the soil surface and replacing it with drier air, which sucks more moisture from the soil. In the hot desert, a raised bed is like a wet sponge laid out on the bare ground. The soil heats up and water evaporates quickly because the sides as well as the top surface of the bed are exposed.

In sunken beds, only the top surface is exposed, and it's partly protected from drying winds by the raised berms. Inside our beds, we dig out the dense clay soil and replace it with at least 18 in. of topsoil, compost and composted manure. Water penetrates quickly into this mixture, and it's retained well by the organic matter. But the berms themselves are the real advantage of sunken beds: They contain rainfall and irrigation water, so none is lost to runoff.

We didn't invent sunken beds—they're common in many dryland areas. We remember seeing little nest-like gardens planted with onions, greens, okra or peppers in West Africa. Sometimes they were very small, just big enough to grow a few plants near the home. In Mexico, we saw bananas and other perennials planted in shallow basins in dooryard gardens. In New Mexico, the Zuni Indians aligned their series of small rectangular basins into "waffle" gardens.

We generally make rectangular beds, about 3½ ft. to 4 ft. wide by 5 ft. long. These are narrow enough that we can straddle them and reach down into the bed, or reach from either side out to the middle. To dig our garden beds, we first remove the topsoil (about 5 in. or so) from the bed and the surrounding area, and set it aside. The next layer of soil is a dense clay, which we dig out and make into firm, sturdy walkways, patting it into shape with our feet.

As we dig deeper into the beds, the clay soil becomes flecked with white bits of caliche (calcium carbonate), and often a rock-hard layer of caliche is not far below. Caliche is common in the soils of this region, and in our backyard it can start as shallow as 8 in. below the soil surface. We've never reached the bottom of the caliche layer, even when digging tree holes up to 5 ft. deep. Using picks and sometimes a heavy iron digging bar, ap-

Photo, top: David A. Cleveland

propriately called a caliche bar in this area, we dig each garden bed at least 18 in. deep.

Digging through caliche is hard work. Friends of ours have rented jackhammers or backhoes to do the job. (To dig those 5-ft.-deep planting holes for trees, we put on eye goggles and use cold chisels, like miners.) Deep digging is worth the effort, though—it really makes a difference in the hot season. We had some plants growing in beds less than 1 ft. deep that needed watering every couple of days in the summer, but 2-ft.-deep beds can go up to a week between waterings.

After removing and discarding the caliche, we refill each bed with a mixture of topsoil, compost and composted manure. Finally, we level the surface, normally by flooding the bed with water, then moving the soil around to fill in the low spots and take off the high spots. Sometimes we just straddle the bed and pull a straight board over the surface to rake it level.

Between crops, we usually add three or four 5-gal. bucketfuls of compost to each bed and mix it into the full depth of the bed. That's a generous amount, but organic matter breaks down quickly in the heat here. We make as much compost as we can, incorporating household trash, garden debris, mesquite leaves and pods, chopped prickly-pear cactus pads, and horse manure. We renew the beds each year or two, thoroughly remixing the soil with a shovel. It's an easy job compared to the initial digging. From time to time, we repair any erosion on the berms with fresh clay.

Watering the beds—By experimentation and observation, we've learned the best way to water the plants in our sunken beds. We keep the soil around newly planted seeds and seedlings moist by planting them in furrows or depressions within the bed and watering only the immediate area. Seedlings may need watering once a day in the hot season, but their roots penetrate only the top few inches of the soil. Wetting the soil deeper than that is a waste of water. For us, four gallons gives a whole bed of seedlings a good soaking.

By the time the seedlings are several inches tall, we fill the soil in around them, so the bed is level again, and begin to water the bed by flood irrigation. We use a hose and bubbler, opening the faucet as far as possible so the water fills the bed much faster than it soaks in. The bubbler breaks the force of the water to keep it from eroding the soil or mulch in the beds. Gardeners in West Africa achieve the same effect by pouring water onto their gardens through bunches of hand-held leaves.

To check if we've added enough water, we take a long stick or an iron bar

In the authors' garden, vegetables and herbs for the cool season include collards, beets and dill.

The desert gardening year

The annual vegetable crops we grow are familiar to gardeners across the United States. During the cooler months from October to March, brassicas and root crops thrive, along with lettuce, peas, and herbs such as parsley, dill, cilantro and fennel. During the rest of the year, we grow corn, beans, squash, melons, tomatoes, peppers, okra, eggplants, amaranths and basil. Some of our crops were selected by Native American gardeners and farmers in this area over a period of hundreds of years. Others grow from seeds we brought back from our travels, and some are standard garden varieties.

Although we garden year round, seasonal differences are important. We follow a planting schedule that helps us make the most of the cool and warm seasons, while avoiding the most difficult gardening weather as much as possible. In early March, after the danger of frost is past, we plant some beds with short-season (60-90 days) corn, string beans and squash. In other beds, we plant long-season crops such as watermelon, and seedlings of chili peppers, eggplants, tomatoes, tomatilloes and basil that were started indoors in early February.

By late May or early June, the corn, beans and squash have produced a harvest, and their garden beds are fallowed. The few beds with long-season crops are heavily mulched, shaded and carefully maintained through the summer. During this time, fruit set stops in many plants because high temperatures disrupt pollen production. But these plants will start producing again after the rains begin, and often continue through November.

In mid-July, when the summer rains start, we plant more beds with corn, beans, cow peas, some squash and okra. The rise in humidity and some rainfall makes it much easier to start and maintain these crops at this time than it would have been if they were started one or two months earlier.

In October and November, when the temperatures begin to drop, we direct-sow seeds of lettuce, greens, peas, carrots and other cool-season crops. We have fun combining plants with different colors, textures and growth forms, such as ruby chard and carrots. The garden looks pretty in the winter, and it's a pleasant time of year to work outdoors.

There may be frosts anytime from the end of November to early March, but the days are sunny and mild. Occasional light rains mean we may not need to water for weeks at a time. The winter crops, and a variety of spring-flowering annual flowers, reach their peak in late March and early April, then decline in the hot, dry weeks that follow. —*D.S. and D.C.*

and insert it two or three places in a bed, feeling how it moves through the soil. With gentle pressure, the stick will move easily in wet soil, slowing or stopping when it reaches a dry zone. If we detect that the soil isn't moist to below the plants' root zones, we add more water and test again. We've learned that flooding with about 3 in. of water at each irrigation will wet the soil to the bottom of the bed. This takes about five minutes with the hose at full blast.

One technique we've experimented with is vertical mulching. A vertical mulch is a column or row of plant stalks buried a foot or so deep in the garden bed, creating a pathway for the water to follow down to the root zone. The idea is to direct the water below the surface layer, where most evaporation occurs. We've tried putting a row of vertical mulch down the middle of a bed, using stalks from last year's corn, Jerusalem artichokes, sunflowers or sorghum. It seems to help speed up water penetration, especially through heavy clay soils. It doesn't make as much difference in porous or open-textured soils, but may still be helpful.

It's important to catch what rainwater we do get, and not lose it as runoff. Our winter rains often are slow and gentle, but our summer rains, or "monsoons," tend to come as short, intense thunderstorms, and ½ in. or more of rain can fall in a sudden burst. A lot of the water runs right over the surface of hard, compacted soil without soaking in—this causes flash flooding in low-lying areas. Right after a rain, our yard is full of water, but it's a thin layer that quickly flows away.

To take advantage of the rainfall, we have to catch it and hold it so it can soak into the soil. We make channels that direct the runoff from the house roof toward basins around the jujube, lemon and pomegranate trees, and we go out with trowels during a rain to make sure that the water is flowing into these wells around the trees, and that the surrounding berms contain it there for a few hours until it can soak in. A heavy rainfall may also flood the sunken beds of our vegetable garden, but there the water sinks quickly into the deep, improved soil in the basins and never stands for more than an hour.

After the soil has been saturated by irrigation or rainfall, surface mulches help retain the moisture by reducing evaporation. They form a barrier to water vapor leaving the soil and they shade the surface, lowering soil temperatures. We use a 3-in.- to 5-in.-deep layer of coarsely textured compost, horse manure, leaf litter, palm fronds or other organic matter to mulch the soil in the planting basins. The mulch should shade the soil between the plants, but around seedlings we try to keep the mulch from touching the plants themselves. It's a haven for

Tender young plants appreciate shading from the hot sun and drying winds. Soleri and Cleveland use carrizo (reed), palm fronds and corn stalks to build sunshades and windscreens.

fungi and sow bugs, both of which prey on tender stems.

Around our mesquites and grapefruit tree, we use a mulch made of closely spaced river cobbles, smoothly rounded stones about 6 in. to 8 in. in diameter. We decided to try this several years ago, when we observed that wildflower and grass seeds were most likely to germinate near bits of gravel, perhaps because the soil was slightly moister under the pebbles. About the same time, we read that stone surface mulches have been used for centuries in arid northwestern China.

Stone mulches are thought to work by capturing the water vapor released in the early morning, even by desert soils. The water condenses on the stones' lower surface and returns to the soil. The stones look nice around the trees—we laid them in pleasing patterns. We don't recommend stone mulches for annuals, however, because the stones absorb and then radiate too much heat for low, tender plants. Also, having to remove the stones each time the soil was prepared and planted would be a lot of work.

Shading—Most of the water plants take up in their roots evaporates from their

leaves. This process, which is called transpiration, helps keep a plant cool and is essential for photosynthesis. Good growth and garden yields depend on sufficient water in the soil for high rates of transpiration and photosynthesis. High temperatures, wind and sun, however, can increase transpiration well beyond what is needed for maximum production.

Shading reduces leaf temperatures, so it cuts down on transpiration. When planting our summer garden, we choose beds that will have the best protection from the scorching midday and afternoon sun. Our garden is interplanted with trees that are relatively tolerant of heat and dryness. They shade the more vulnerable vegetable plants. For example, we grow squash and melons under a big mesquite tree, and train the vines back into the shelter under the tree, not out across the burning hot soil. Tall annuals such as corn, sorghum and sunflowers can give shade and protection to shorter plants. The sun is so intense here that vegetables get all the light they need even under partial shade.

Providing shade for small seedlings is especially important. We use the tips of palm fronds or make a tepee of corn stalks around tender young plants. We usually orient the shades to protect the plants on the south and west sides, shielding them from the worst of the midday and afternoon heat.

We also construct shades over the garden beds in the summer, using the woody stems of carrizo (Phragmites australis), a native reed, and palm fronds. These building materials are attractive and easy to grow or find in this area. They make excellent, lightweight, flexible shades that won't break or blow over with the winds of late summer. For vining varieties of tomatoes, beans and tomatilloes, these shades also serve as trellises. We stick the poles about a foot deep into the soil (this is easiest to do after watering), then tie on crossbars, tops or sides as needed. We often line the structures with palm fronds, so they serve as windbreaks as well as sunshades.

No matter how adapted the crop or how carefully managed the garden, when there is no rain the summer desert garden needs to be watered to ensure that the plants will survive. In the summertime, we always try to water our garden in the evening. During the night, as the air cools (and with the absence of sunlight), evaporation and transpiration rates drop, so less water is lost to the air. Warm desert evenings are wonderful, the most pleasant time of the day, and we relax and enjoy the garden while avoiding the heat and sun. □

Daniela Soleri and David Cleveland are directors of the Center for People, Food and Environment in Tucson, Arizona.

Reclaiming a Lifeless Soil

A summer's work makes a difference

by Mark Trela

Soil horror stories are a dime a dozen. Many gardeners, serious or casual, have faced a backyard compacted into a clay tennis court, or have tried to work a soil as barren as the surface of the moon. Hard-packed soil is a common legacy left by heavy equipment at the site of a newly constructed or renovated house, but it could just as easily result from soil mismanagement or lack of tending. What follows is a true-to-life soil horror story; only the soil has been changed.

For the past three years, I've worked as a gardener at the New Harmony Inn in New Harmony, Indiana, site of an early 19th-century experimental utopian community, and presently a peaceful vacation retreat and meeting place. Under the direction of Mrs. Jane B. Owen, also a devoted gardener, the existing inn was expanded and a conference center added. The 1800-sq.-ft. area between the two was

planned as a cloister garden, an enclosed space traditionally used for growing greens and herbs in medieval monasteries. Brick walkways would lead visitors through the garden.

The site was perfect for a cloister. Bounded by high brick walls, the area was accessible only by mounting seven steps in one corner. But it was a gardener's nightmare. During construction, the cloister area had been excavated to a depth of 3 ft. and had served as a driveway for heavy equipment, resulting in compacted soil. Once the buildings had been completed, rust-colored, gritty-clay fill dirt was trucked in to fill the hole, and 6 in. of "topsoil" was smeared onto the fill dirt, only to be removed later by the contractor to correct the grade. Concrete-mixing trucks frequently hosed excess concrete onto the ground. In anticipation of this soil abuse, I had removed and stockpiled about five tons of fertile topsoil from the site before construction began, intending to rebuild the cloister soil with this reserve later on. But by the time the garden area was ready to be worked, this soil had been used for other gar-

dening projects. The heavily compacted fill dirt and underlying clay subsoil, all seemingly devoid of organic matter, were my inheritance.

When confronted with the cloister-garden fill, I was immediately tempted to dial 1-800-TOP-SOIL for a quick, trucked-in fix, but that would have cost a fortune. In my frustration, I almost ignored my previous four years of training and practice in biodynamic gardening, experience that had convinced me that a lifeless soil can be transformed over a period of time. The biodynamic gardening method, which originated in Germany in the early 20th century, is a holistic approach to farming and gardening. In simplest terms, biodynamics is concerned with creating and maintaining fertile soils as the foundation for healthy plant growth. Rather than by direct fertilization of plants, friable, nutrient-rich soil is

As part of his program to improve the soil in this cloister garden (above), Trela planted a buckwheat green manure. Its poor performance called attention to an unanticipated drainage problem, which he solved by installing plastic drainpipe.

developed by the incorporation of organic material such as compost and green manures (crops grown on and tilled into a site, to provide organic matter). Various biodynamic preparations (whose composition is a closely guarded secret) often are added to stimulate the microbial and other biological activity necessary for transforming organic matter into fertile soil. This garden would be a severe test of these beliefs and techniques. Fortunately, Mrs. Owen supported my approach, and we embarked on a reclamation effort in mid-March 1987.

My plan was to add three or four cubic yards of compost, then plant and till-in summer, fall and winter green-manure crops. But rarely do plans and execution coincide. I didn't have nearly enough finished compost, so I had to rely on green manures. And when I was ready to begin, I found out I had just two weeks to make the cloister presentable for a formal dedication of the new facility. Armed only with shovels, my helper, Suzanne Davoust, and I began to spade the entire 1600-sq.-ft. planting area 8 in. deep. The ground resisted, yielding nothing but wet, shovel-shaped clods that didn't readily break up. The soil was extremely wet in some spots, which I dismissed at the time as a consequence of moist March weather, although it foreshadowed later problems. Because of the excessive moisture, I decided not to till the soil right away, which might risk more compaction.

A few days before the dedication ceremony, the soil was still too wet to till or plant, so I mulched the garden with shredded leaves to give it a uniform, more pleasing appearance. I had a plentiful supply of partially decomposed hard-maple, hackberry and elm leaves (I gather leaves each fall). Suzanne and I dragged about 20 large garden carts of leaf mulch up the stairs into the cloister, an interesting exercise that I hope will not be repeated soon. We shredded the leaves to increase the surface area available for bacterial decomposers, and spread them about 3 in. thick.

Many organic gardeners till leaves into the soil in the fall, letting them decompose during the winter. Although leaves tilled-in during the spring might temporarily tie up nitrogen and acidify soil as they decompose, I was committed to using amendments already at hand or ones I could grow, rather than importing them. Also, I like to experiment, and I felt I would have an advantage because I applied Bio-Dynamic Field Spray (available from Pfeiffer Foundation, Inc., Threefold Farm, Spring Valley, NY 10977) when tilling-in the leaves. The field spray, a powdery, aromatic dust that contains soil bacteria and biodynamic preparations, is designed to speed up decomposition

Buckwheat flourished once the drainage had been improved. Six-week-old plants were tilled under as a green-manure crop.

Royalty Purple Pod beans flank straw-mulched emerging buckwheat in the large cloister-garden bed, providing another source of green manure for rejuvenating the soil.

of leaves and green manures.

Along with the leaves, I added 200 lb. of gypsum (calcium sulfate), a soil conditioner I routinely add to compacted, heavy clays, hoping it would perform as advertised: "Millions of tiny hoes breaking up the clay." My application rate was a compromise between the rates recommended for annuals and for lawns. (I prefer to apply gypsum along with compost in the fall so that it has more time to act before I plant, but once again time limitations required immediate action.) About ten days later, when the soil had dried out enough, I tilled, setting my tiller so that it churned up just the top 4 in. of soil, in order to avoid burying the leaves too deeply and creating an anaerobic mess in the clay.

I turned to buckwheat (*Fagopyrum esculentum*) after the leaves, gypsum and field spray had been incorporated, with great hopes that this versatile green-manure crop would become the heroine of my soil drama. I had previously experimented with buckwheat and liked what I saw: an unassuming plant that grew rapidly, had a short life cycle, was adaptable to

depleted soils and dry conditions, and had a reputation as a potassium accumulator.

I planted the first buckwheat crop on April 13, about two weeks after tilling, when the leaves were sufficiently broken down. Although buckwheat grows best in warmer conditions than our typical mid-April temperatures, an unusually warm week with 70°F temperatures preceded planting. I broadcast about 1 lb. of the pyramid-shaped seeds and raked them into the soil surface, sowing them thickly to get a dense stand, which never seems to stress this species. The seedlings, with their distinctive trifoliate cotyledons, emerged in a few days. I kept the soil moist until the seeds had germinated, but after the plants had grown past the cotyledon stage, I let it dry out between irrigations. Once the stand had filled in, it shielded the soil from the drying effects of the sun and wind. I didn't have bird problems, but if I had, a 1-in. or 2-in. straw cover would have prevented them from feasting on the seeds.

For the next month, the buckwheat plants were stunted, and by mid-May the second and third sets of true leaves had

Photos: top, Mark Trela; bottom, Staff

yellow leaf margins, signaling stress. At first I thought the decomposing leaves and gypsum were responsible, but after digging a few deep holes in the soil, I came to a much different conclusion. Eight inches down, I discovered sodden, compacted soil covered with a thin sheet of water and shriveled buckwheat roots, which had stopped growing when they reached this anaerobic layer.

It was obvious that part of the soil's problem, at least, was caused by poor drainage. By this time, Mrs. Owen had decided to plant grass rather than herbs in the center 800-sq.-ft. bed, so we concentrated our drainage efforts there. Before tackling the drainage installation, we applied field spray and tilled the buckwheat under on May 22. This time I tilled with a Mantis tiller (available from Mantis Manufacturing Co., 1458 County Line Rd., Huntingdon Valley, PA 19006), instead of the big tiller I had previously used. Its small tines revolve four times faster than those of a large tiller, which made it ideal for breaking up clods and incorporating the buckwheat.

After tilling, my comrade and I grabbed our shovels and spent one gloriously humid afternoon digging three 10-in.-wide trenches 10 ft. apart in the garden, sloping them toward an existing drain. We laid 4-in.-dia., flexible, perforated plastic drainpipe in the trenches on a thin layer of pea gravel, capping the uphill end of each pipe with a double layer of 6-mil plastic held in place with a stout rubber band. Then we covered the drainlines with a ton of the gravel, which acts as a coarse filter and prevents soil from blocking drainage slits in the pipe, and backfilled along the sides and top of the pipe with the gravel and soil.

While excavating the trenches, we noticed slight changes in the color, structure and odor of the top few inches of soil compared to the condition we had found when we first spaded it in March. Some parts of the upper 4 in. of soil were dark brown instead of rusty orange, crumbled readily in our hands, and released a faint, pleasant odor, while the lower layers were still rust-colored and odorless. Nothing dramatic, but we felt it was a subtle sign of positive changes to come. As we dug, we saved this improved layer and hauled away a truckful of the lifeless fill soil.

I sowed buckwheat again in the center bed and anxiously awaited results. The seedlings emerged after five days. After four weeks, the 1-ft.-tall plants were thriving, with none of the telltale stress signs we had seen in the first planting. We tilled-in the crop about two weeks later. Next I decided to plant Royalty Purple Pod beans as a green manure. Although buckwheat adds organic matter to the soil, Purple Pod beans have extra advantages—they fix nitrogen, and are attractive to the eye and palate. I mixed the seeds with a legume innoculant prior to planting, figuring that the rhizobia bacteria responsible for nitrogen fixation would be absent in my depleted soil.

Because I ran out of bean seeds, I planted buckwheat in a 10-ft.-wide swath down the center of the garden, using a straw mulch to keep the soil moist. To my great satisfaction, both species grew rapidly and the beans produced a modest but tasty crop. I unearthed some of the bean plants about one month after planting and found some nitrogen-fixing nodules on the roots, although they were not in the profusion I've seen on bean roots. Beetles chewed some of the leaves, but nothing fatal. We bid the beans and buckwheat a fond good-bye when we tilled them under at the end of August. I added about a ton of well-ripened compost from a pile I had started in April, and then let the garden rest unplanted, but watered, until it was cool enough to lay sod at the end of September.

In six months' time, I was gratified to see small but significant changes in the cloister-garden soil. Most striking was the development of a rich, brown color and a crumblier structure. No more clods in the shape of a shovel when I spaded. A small band of hearty earthworms had taken up residence, though it was by no means a mass migration. I usually don't test my soil, but as I wrote this article, I was curious as to whether the changes I had observed would show up in a soil test. So I sent samples of the improved soil and of the unimproved fill dirt still remaining on the garden's periphery to Woods End Laboratory in Mt. Vernon, Maine, for analysis. A summary of the results is shown at the top of p. 88.

Having had only one season to build the soil in the cloister garden, I find it difficult to make generalizations for other gardeners faced with a bleak soil situation. Soil improvement is an arduous and ongoing process that might not be everyone's cup of tea, but I think my efforts were worthwhile. In the absence of compost, the introduction of leaves, green manure or other organic matter into the soil creates a hospitable environment for microorganisms and other soil creatures, and begins a cycle of soil rejuvenation. I learned some things, too. I've always known that buckwheat is an effective soil-improver, but in the past I've used it to prepare for vegetable gardens. Now I realize buckwheat can also be grown as an attractive, temporary ground cover to improve the compacted, nutrient-depleted soil of a future lawn site. ☐

Mark Trela is a biodynamic gardener in New Harmony, Indiana.

Notes from a soil scientist
by Bob Dahse

Like Mark Trela, many gardeners face the task of making a healthy topsoil from a lifeless heap of backfill. Soil tests can be an important tool in soil improvement— an initial test helps you figure out the nutrient levels you're starting with, and later tests indicate any changes. In a healthy soil, a test may provide only a snapshot of the rapidly changing nutrient levels resulting from vibrant bacterial activity. But in biologically inactive fill soils like Trela's, soil tests give a very stable (if sometimes depressing) representation of mineral levels and their relative balance, and an accurate idea of what's needed to restore a mineral balance favorable for growth of microbes, worms and, eventually, crops.

Sometimes the test results are baffling unless you've spent some time learning about soil-component levels. It's not often apparent from the numbers and lab recommendations how your cultural practices might affect soil health. A basic soil analysis, such as Trela's, typically tests for phosphorus and potassium (macronutrients used in large quantities by plants), and for calcium and magnesium (also required for plant growth, but in smaller amounts). Sometimes soil nitrogen is estimated, or one of the several forms of nitrogen present in soil is measured, but the values arrived at don't give a meaningful picture of total nitrogen or its availability. Labs usually test to determine pH, soil texture (the proportion of sand, silt and clay) and the percentage of organic matter in the soil, all of which influence nutrient supply and availability. I looked at Trela's soil-test results and his cultural practices, and here are some thoughts from a soil scientist's point of view about what might have happened in his garden and why changes occurred.

The test of the unimproved fill soil certainly looks unpromising for growing healthy plants. High pH limits availability of phosphorus, and of trace minerals such as iron, manganese and zinc. A pH range of 5.5 to 7.9 is considered optimum for the crops Trela had planned to grow. The low percentage of organic matter— 3% would be desirable for Trela's predominantly clay soil—indicates limited nutrient availability, minimal biological activity and poor soil structure. Excessive calcium contributes to the concrete-like structure of the soil. Calcium was high in relation to magnesium and potassium. This imbalance interferes with a plant's ability to take up magnesium and potassium, and contributes to stunted plant growth. A desirable ratio of calcium

SOIL REPORT	Soil pH	% organic content	Texture	Nutrients in lb./acre			
				Available phosphorus	Calcium	Magnesium	Potassium
Unimproved soil	8.1 (in water)	0.3	Clay	57	12,100	490	120
Improved soil	7.8 (in water)	1.8	Clay	88	8,100	410	220

to magnesium to potassium for the herbs and vegetables Trela had originally intended to grow would be approximately 10:2:1.

Trela's fill soil was not typical of subsoils used for backfill in the Midwest to the East Coast. Few subsoils have such a high pH and high calcium content, although they may share a low organic-matter content. Typical subsoils in these areas can have low levels of calcium, magnesium, potassium or phosphorus, and are quite acidic (low pH); the topsoil rarely comes close to a neutral pH of 7.

Even without the benefit of an initial soil test, Trela's approach was a good one. The changes he observed in the color, odor and structure of the soil reflect a successful alteration in soil chemistry and biology. The final soil test leaves no doubt about the improvement. The percentage of organic matter had increased sixfold, the calcium level was reduced by one-third, and the calcium:magnesium: potassium balance had improved.

No single process or product Trela employed improved the soil. In simplest terms, soil building requires vast numbers of a variety of organisms—bacteria and fungi are the chief agents—and organic material and minerals for them to eat. By incorporating leaves, green manures, gypsum and biodynamic field spray, Trela supplied the microbes with organic carbon, nitrogen and sulfur, all of which are required for their growth and reproduction. Better aeration and drainage provided oxygen, which stimulated root growth of the green-manure crops. Incorporating the green manures into the soil recycled easily leached, water-soluble nutrients, and supplied even more nitrogen and organic carbon to the microorganisms, improving soil structure and nutrient capacity. Improved drainage also provided a hospitable environment for earthworms. All of these efforts worked in concert to stimulate biological activity. The net result was a soil with less compaction, more available nutrients and more organic matter.

Let's look more closely at Trela's procedures and their consequences. Spading the top 8 in. of soil in an attempt to physically loosen the cloddy clay soil may have had mixed results. As Trela later discovered, working a soil with high clay

content while it was wet probably contributed to an already compacted, poorly drained "pan" layer 8 in. down, exactly the depth the shovel reached. The yellow leaf edges on the first buckwheat planting were symptomatic of a potassium deficiency. They were ample testimony that even a reputed potassium accumulator like buckwheat can't obtain adequate potassium through oxygen-starved roots dangling in stagnant water. But even this stunted planting added organic matter and encouraged biological activity in the top few inches of the soil when it was tilled-in. And once Trela had improved the drainage, the second planting of buckwheat and beans grew better and contributed more organic matter.

Adding gypsum (calcium sulfate) provided soil bacteria with quickly available sulfur, which, along with nitrogen and carbon in the buckwheat and tree leaves, helped complete the microbes' diet. As bacterial populations multiplied, they produced a slimy substance that coated the fine clay soil particles and caused them to clump together, or aggregate. Such aggregated particles are separated by large pores, resulting in less compacted, more friable soil. If Trela had tested the soil at the beginning of the season, and thus been aware of the high calcium level, instead of gypsum he could have added a sulfur source that didn't contain calcium, such as sul-po-mag (a blend of potassium and magnesium sulfates), along with elemental sulfur. This combination, plus actively decaying organic material, would have increased the potassium and magnesium levels, bringing them more in balance with the excessive calcium, and would have decreased the pH more than sul-po-mag alone could. Elemental sulfur slowly releases sulfates over a three-year period, and would continue to reduce the pH once sod was planted.

The tests show a huge increase in organic matter—the equivalent of adding 800 lb. of dry organic debris to the garden. Only part of this came from the leaves and green manure; the rest was created primarily from existing raw materials (free mineral carbonates) already in the soil. Increased bacterial populations hastened the decomposition of organic matter from all sources and

the formation of humus. These processes released weak organic acids that slightly reduced the pH and broke down the stable salt, calcium phosphate, making more phosphorus available to the plants. Decomposition of organic matter also supplied carbohydrates, which further stimulated bacterial growth. The soil color darkened as humus formed, and microorganisms called actinomycetes produced a fragrant odor.

As the bacteria digested the green manure, and to a lesser extent the tree leaves, they produced, among other things, ammonia, which in turn triggered several changes in the cloister garden. The ammonia provided a nitrogen source for nitrate- and nitrite-producing bacteria. It reacted with water to form hydroxyl ions that oxidized the humus, releasing carbon that further fed the bacteria. Ammonium, which is also formed when ammonia reacts with water in the soil, bumps potassium off storage sites on clay particles, increasing the amount of potassium available to plants.

One of the most obvious effects of ammonia was the decrease in calcium. When the cement sweepings (calcium carbonate) and gypsum (calcium sulfate) were added to the moist soil, positively charged calcium ions were released, and these attached themselves to the negatively charged surfaces of the clay particles. Ammonium, though, substitutes for calcium stored on the clay. The released calcium ions bind with free nitrates, forming calcium nitrate, which is easily leached away by water. So even though Trela had added calcium, in the form of gypsum, he ended up with reduced calcium levels, resulting in a more balanced calcium:magnesium: potassium ratio and better plant growth.

With all the soil improvements, will the sod crop thrive? Without further work to balance mineral levels, lower the soil pH and stimulate humus production, the sod may suffer from nutrient shortages, despite the thin topsoil layer imported with it and the addition of compost before planting. Improvements in a nearly dead soil can be dramatic, but a healthy, self-sustaining soil requires a commitment to a long-term relationship. □

Bob Dahse is a soil analyst and consultant in Winona, Minnesota.

No More Lawn-Mower Bag

Grass clippings nourish the lawn and don't cause thatch

by William E. Knoop

For years most of us have mowed our lawns, bagged the clippings and given them away. The city truck took the bags, and we forgot about them.

It's time to change. For one thing, many municipalities across the country have banned yard waste, trying to eke a few more years from their brimming landfills. And grass clippings are a big item. By volume, they average 20% to 25% of the waste in many towns. So now homeowners are faced with the prospect of dealing with all those clippings on their own. There is a simple solution: don't do anything.

The best way to deal with grass clippings is to let them fall on the lawn and decompose. Grass clippings are rich in nutrients, so recycling them nourishes the lawn. In fact, grass clippings contain all the nutrients essential for plant growth, and in amounts equal to what is found in many of the so-called natural organic fertilizers on the market.

What's more, clippings are quick to break down. Because clippings have a very high water content, they wither quickly and shrink enough to sift down through the lawn to the ground, where fungi and bacteria attack them rapidly. Once the soil organisms have completed their work, the nutrients in the clippings become available to the lawn.

If you stop bagging grass clippings, you'll be in good company. Professional turf managers have always considered grass clippings too valuable to

Throwing clippings onto a new-mown lawn, a side-discharge mower leaves barely visible windrows. The clippings will soon sift to the ground and decompose, fertilizing the lawn.

All photos: Mark Kane

throw away. The people who manage golf courses, parks, cemeteries and athletic fields have never removed the clippings.

The thatch myth

One of the most common misunderstandings about leaving clippings on the lawn is the belief that they create thatch. This is simply not true. Too much nitrogen fertilizer and over-watering are the usual causes of thatch.

Thatch is a layer of living and dead plant tissue on the soil surface. When thatch builds up, it can block water and air from the soil below. Because thatch superficially looks like dead grass clippings, the two have become associated in the public's mind. But they are not connected. Thatch is composed largely of grass roots, crowns and rhizomes and stolons (underground or creeping stems). Because these parts are very high in lignin, a cell wall material that does not decompose readily, they can build up. Grass leaves, however, contain only a small amount of lignin.

Change two habits

In a program for homeowners called "Don't Bag It," my fellow turf grass specialists of the Agricultural Extension Service at Texas A&M University recommend two basic changes of habit: mowing more often and using slow-release fertilizer.

Mow more often—Mowing every five or six days, instead of weekly, is the first change. People bag clippings because they don't like to look at the browning grass blades for the day or two they take to disappear. Mowing slightly more frequently helps solve that problem. Ideally a lawn would be mowed five times a month instead of four. Although you'll have to mow more often, you'll find the job takes 40% less time because there's no bag to bother with. When the lawn grows slowly, you can mow less frequently. The rule of thumb is never cut more than one-third of the lawn height.

When you mow, the clippings can be discharged in several ways. If you simply remove the bag, a side-discharge mower will leave the clippings in rows. There's no problem with leaving the rows, but most people object to their appearance. It's better to close the door on the chute. The mower will spread the clippings more evenly, and, incidentally, cut them even smaller. If the chute on your mower lacks a door, ask a lawn equipment sales or service dealer about attaching one.

Before long, new mowers will have no discharge chute and no bagging attachment. Manufacturers are now making what they call mulching mowers that circulate the clippings inside the mower deck, cutting them repeatedly into smaller pieces, which disappear more quickly after they fall on the lawn. But if you have an older mower, don't worry. The clippings from an ordinary mower will decompose just fine. Many manufacturers also offer kits to convert old mowers to mulching ones.

Change fertilizers—If you want to stop bagging, you also have to use a fertilizer that has at least half of its nitrogen in a slowly soluble or slow-release form. A lawn needs a small amount of nitrogen every day of the growing season. Look for a fertilizer that says it's slow-release, slowly available or controlled-release. The label may also say "water-insoluble nitrogen." Avoid fertilizers that have lots of highly soluble nitrogen. They release the nitrogen too quickly, causing the plant to grow leaves very rapidly, at the expense of healthy stem and root growth. Then you have to mow more frequently.

Another good rule is never to apply more than one pound of nitrogen per 1,000 square feet at any one time. The first of the three numbers on a bag of fertilizer indicates the percentage of nitrogen. For example, a 15-5-10 fertilizer is 15% nitrogen. Because 15% means one-sixth of each pound of fertilizer is nitrogen, you would spread six pounds of fertilizer to apply one pound of nitrogen.

Finally, fertilize the lawn only when the grass is actively growing. In the South this means from late spring to early fall. In the North the main active growth periods are spring and fall with slower growth during the hot summer. Your local Cooperative Extension Office can give you exact dates.

"Don't Bag It" began four years ago and now has thousands of satisfied ex-baggers all over the country. It's a simple program: stop bagging, mow a little more often, and use a suitable fertilizer. If you follow the program, your lawn will do much better.   □

William E. Knoop is a turf grass specialist with the Texas Agricultural Extension Service in Dallas, Texas.

Professional turf managers have always considered grass clippings too valuable to throw away.

Grass clippings shrink dramatically as they dry and soon disappear into the lawn. The pile on the right is a coffee-can's worth of fresh grass clippings. The pile on the left started at the same size, but has dried for five days. Because they were stored out of the sun, the clippings kept their color as they dried.

Houseplant Hydroponics

Soilless system in a pot

by Melitta Z. Collier

Inside hydroponics
Any kind of houseplant, such as this spathiphyllum, can grow without soil. The pot is filled with fired-clay stones and watered with a nutrient solution.

Water-level indicator

1-in.-dia. Styrofoam dowel

Moisture content of stones ranges from 100% at bottom to dry at top.

Roots grow in stones, not in water.

1¼-in.-dia. PVC pipe

Cut four slots in base of pipe.

Water level shouldn't exceed one-third height of pot.

Healthy potted plants add beauty and grace to any room, but some people find it hard to keep plants looking their best. Overwatering, underwatering, over- or underfertilizing—many gardeners just can't seem to get it right. Often the problem is in the potting mix. What's needed is something that will drain well, stay evenly moist, provide nourishment and support the plant. The answer I've discovered is hydroponics—growing plants without soil.

Hydroponic systems for food crops can be highly sophisticated, but my interest is decorative plants, and I use a very simple system of man-made clay stones and commercial fertilizer in plastic pots with water-level indicators. The same setup works fine for all the houseplants I grow—foliage plants, orchids, African violets, begonias, ferns, palms and even cacti.

Though the "hydro" in hydroponics means water, that doesn't mean that the plant's roots are immersed in water. At maximum, the water level shouldn't exceed one-third the depth of the container. The roots grow in the stones, which draw up moisture from the bottom of the pot by capillary action. Because the degree of moisture in the stones ranges from 100% at the bottom of the pot to almost none at the top, a plant's roots can develop at whatever moisture level they require. There is constant, even moisture but never sogginess. And because the stones never break down or become compacted, air can circulate around the roots at all times.

Compared to growing plants in potting soil, hydroponics is a breeze. The stones are clean, light and odorless, and soil-borne insects and fungi can't live in them. The plants can go for days or weeks without attention, which is ideal for people who travel often. And you never have to worry about when or how much to water—anyone can look at a water gauge and tell whether water is needed.

The materials

Stones—I've been growing plants in clay stones for about 15 years. The stones are roughly the size of marbles, and have bricklike color and texture. They're available from hydroponics suppliers around the country or by mail order (see Resources on p. 92). A 20-lb. bag costs about $15. Stones last indefinitely, and you can reuse them to pot different plants. Whenever I empty out a pot, I wash the stones in clear water and save them. Before reusing them, I put them in a pan with ½ in. of water in the bottom and heat them in the microwave for

Illustration: Frances B. Ashforth

four minutes to sterilize them.

I think it's worthwhile to invest in these manufactured clay stones, because they work better than anything else I've tried so far. People often ask me if they could make their own stones by breaking up bricks or clay flower pots, but I don't recommend it. The debris would have too much dust and grit, which packs down, holds too much moisture and doesn't let enough air penetrate around the roots. Likewise, coarse sand, chicken grit (sold at feed stores) and perlite are all too fine-textured to work well. Recently I've been experimenting with a kind of heat-treated slate, used by ready-mix concrete manufacturers as a lightweight alternative to gravel. The slate "stones" are gray, and come in small, medium or large grades. I'm trying the small and medium sizes, and so far they seem to work fine.

Fertilizer—Unlike potting soil, clay stones are inert and don't supply any minerals. Instead, carefully measured doses of nutrients must be added to the water. The "water" in a hydroponic system is actually a dilute fertilizer solution that provides a constant supply of dissolved nutrients to the plants. What's important is that the fertilizer include all the needed micronutrients and trace elements. (Fertilizers designed for use with soil often supply only the macronutrients—nitrogen, phosphorus and potassium—not the entire palette.)

Several brands of fertilizer are available through hydroponic suppliers. For years I used a liquid 2-2-2 formula that I mixed with the water each time I filled the containers, but it's troublesome to mix enough fertilizer solution to water as many plants as I have. I recently switched to a granular timed-release fertilizer, which I sprinkle over the stones every four to six months, and I add plain water in between times.

Hydropots—Pots designed for hydroponic growing come in a great variety of sizes, shapes and colors, but they all have one thing in common: water-level indicators. One type of indicator is a window in the water reservoir at the bottom of the two-part pot. The other is a tube with a float that rises and falls with the water level in the pot. Both tell you exactly when and how much to water.

You can also make your own hydropots, using containers made of any *nonporous* material such as plastic, glass or glazed ceramic. Size, shape and proportions don't matter—the pot can be short and wide or tall and skinny, square, round, or whatever. If there are drainage holes in the bottom, use florists' clay to plug them. Press the clay firmly into place, then fill the pot with several inches of water and let it stand overnight to be certain you have a good seal.

For inexpensive large containers, I use plastic buckets or waste baskets, and make water-level gauges from Styrofoam dowels that float inside PVC pipe uprights. I cut the pipe the same height as the pot, and the dowel two-thirds as tall. Installed in the hydropot, the dowel rises to the top of the pipe when I've added enough water, and recedes as the water is used up. In addition, I drill a single small hole through the plastic pot about one-third of the way up from the bottom. That's so rain won't overfill the pots when I take plants outside for the summer. Come fall, I plug the holes with florists' clay when I bring the plants back indoors, so water can't spill on the carpet.

Planting a hydropot

You can grow any plant you choose in hydroculture. One approach is to take a plant that's already growing in a potting soil and transfer it. The photos on the facing page show how to do this.

Rooted cuttings are also great candidates for hydroponic culture. My favorite method produces strong root systems ready for the hydroponic system and requires very little monitoring during the process. I root cuttings in covered, clear-plastic shoe or bread boxes, or in an old fish tank covered with a piece of glass. I use perlite as a rooting medium because it holds moisture well and normally doesn't have to be watered again before the cuttings have rooted. Before hydropotting a rooted cutting, gently run water through the roots to wash out most of the perlite. (It's okay if a little perlite clings to the roots, because unlike soil, perlite doesn't break down.) Sharp sand can also be used, but it dries out faster, must be watched more carefully, and usually has to be watered several times before the roots have developed.

Hydro maintenance

Maintaining plants in hydropots is easy. The water level in the container is important. It should never be more than one-third the depth of the pot. Looking at the water-level indicator will tell you if it's time to water. When the window is empty or the float has dropped to the lowest point, simply pour water through the stones until the window is full or the float rises to the top. Let the plant use up the water before you refill; don't keep "topping off" the water. Depending on the size of the pot and plant, the time between waterings can be from a week or so to a month or more.

Leaching—This is a chore that's often neglected with soil-grown plants, because it's messy. The purpose is to flush out the mineral salts (from tap water and fertilizer) that collect in the pot and on the roots. In hydroponic growing, it should be done every six months or so, or whenever a white haze is noticeable on the stones. The large potted plants that I set outdoors for the summer are leached when excess rainwater flows through the pots and drains out the hole in the side of each

RESOURCES

Hydroponic Society of America, c/o Gene Brisbon, P.O. Box 6067, Concord, CA 94524. 415-682-4193. Annual membership is $25.00. The society publishes six newsletters a year, sells books, and sponsors conventions and tours.

Applied Hydroponics—Eastern U.S., 208 Route 13, Bristol, PA 19007. 800-227-4567. Catalog free.

Applied Hydroponics—Western U.S., 3135 Kerner Blvd., San Rafael, CA 94901. 800-634-9999. Catalog free.

JC's Garden Center, 9915 S.E. Foster Rd., Portland, OR 97266. 800-233-5729. Catalog free.

Leni Hydro-Kultur, 4420 Parkway Commerce Blvd., Orlando, FL 32808. 407-295-8567. Call or write for retail price list or the location of the dealer nearest you.

Author Collier grows orchids, begonias, ferns, palms and other houseplants hydroponically. The plants thrive, and all Collier has to do is replenish the nutrient solution periodically.

Transferring a plant from soil to hydroponics

Let the soil dry out as much as possible (but not so much that the plant begins to wilt), and slide the root ball out of the pot. With a thoroughly root-bound plant, cut off the bottom third of the soil and root mass. This will make it easier to remove the soil lodged at the center. Besides, the plant won't need such an extensive root system when it's in hydroponic culture.

Shake off the loose soil, then use a strong spray of water to wash off as much as possible of the soil that clings to the roots. Pick away any remaining bits by hand. It's important to remove *all* the soil, because any that's left behind will decompose and can cause roots to rot. Comb through the roots with your fingers to straighten and separate them. Trim off the tips of some of the longer roots to stimulate the growth of new feeder roots nearer the base of the plant.

Fill the hydropot a little more than one-third full with washed stones. Carefully spread the roots over the stones. Then fill the pot to the top with stones. As you do this, gently tap the pot so the stones will settle in around the roots. —*M.Z.C.*

container. For smaller plants that stay indoors all year, I set each pot in the kitchen sink and run tepid water through the stones for a few minutes. After leaching, I renew the nutrient solution.

Repotting—In the ideal conditions provided by hydroponics, a plant's roots don't have to grow far and wide in search of moisture and nutrients. All the energy that would be used to develop a large root system goes instead to the top growth. This means that plants in hydroponics often grow faster and larger than they would in soil culture, but they don't need repotting as often.

When you have to water more frequently than you choose to, or when a plant has gotten so large that it looks out of proportion with the pot, it's time to shift to the next larger size. Some of the stones will fall away when you take the plant out of the pot, but it isn't necessary to remove the stones embedded in the root system. Prune off any roots that are going around in circles and any that are bruised or damaged. Fill the bottom third of the larger container with stones, set the plant in and fill in with more stones as needed. Because the roots are scarcely disturbed in this repotting, the plant will suffer little if any shock, and you can give it water/nutrient solution immediately.

Plant problems—Hydroponics isn't a cure-all. It won't revive plants that are suffering from pests, diseases, low light, improper temperatures, old age or other ailments. People are always trying to give me their failures, but I don't have enough time and space to play Florence Nightingale to sick plants. Nor do I try to bring coleus, impatiens or other plants in from the garden at the end of summer—they never look as good indoors as they do outside. Instead, I specialize in plants that are comfortable at room temperature, tolerate shade or low light conditions, don't need frequent pruning or grooming, and don't attract insect pests.

The above-ground parts of a plant need the same kinds of care and attention in hydroculture as in soil culture. I shower my plants' leaves with tepid to warm water or wipe them off with a damp cloth. A clean plant looks better and grows better, and regular washing also helps control red spider mites, aphids, mealybugs and other pests. I keep a fan running all the time to circulate the air in my plant room, and have a built-in pool and fountain that increase the humidity. In the summer, I move nearly all my plants outdoors to a lath house, where they're shaded from the sun but open to the rain and air. When I bring them back indoors in the fall, they look fresh and healthy, ready to give me pleasure all winter. □

Melitta Z. Collier lives in Silver Spring, MD.

Index

A

Arid-land gardening:
    crops for, 83-84
    sunken beds for, 80-84

B

Bark chips:
    as mulch, 67
    as soilless mix ingredient, 24
Beans (*Phaseolus* spp.), as green
    manure, 86, 87
Beaubaire, Nancy, on choosing
    fertilizers, 52-55
Beds:
    peat moss in, 72
    raised, for improved soil drainage, 28
    starting, 68-70, 74-75
    sunken, system for, 80-84
Biodynamic gardening, discussed, 85-86
Bir, Richard E., on soil drainage, 26-28
Bonsall, Will, on hot compost, 36-39
Borders, newspaper/mulch creation of,
    70, 75
Buchanan, Rita, on evaluating potting
    soil, 21-25
Buckwheat (*Fagopyrum esculentum*), as
    green manure, 86-87

C

Carney, Nancy, on no-dig beds, 68-70
Cleveland, David A. (joint author). *See*
    Soleri, Daniela.
Cline, Steven, on composting, 32-35
Clover, as ground cover and compost,
    44
Collier, Melitta Z., on hydroponics, 91-93
Compost:
    bins for, 30, 33, 37-38, 75
    hot, making, 36-39
    ingredients for, 33-34
    leaf, using, 31
    making, 32-35, 75
    nutrient content of, 9
    sheet, with leaves, 86
Composting:
    books on, 34
    cold, process of, 9
    hot,
        vs. cold, 30-31
        process of, 9
    leaf, 29-31
    process of, 34-35, 39
Container gardening, repotting for, 93
Containers, for hydroponics, 92

Cover crops:

    selecting and planting, 41-42
    soil building with, 40-41
    sources for, 43
    for spring planting, 42-43
    for summer planting, 43
    turning under, 42
    winter-hardy, 43

D

Dahse, Bob, on soil analyses, 87-88
Deserts. *See* Arid-land gardening.
Drainage, soil:
    importance of, 26-27
    improving, 28, 87
    test for, 26

E

Earthworms, as beneficial soil organisms,
    14

F

Fertilizer:
    buyer's guide to, 52-55
    determining need for, 53
    for hydroponics, 92
    labels of, reading, 54-55
    organic vs. synthetic, 55
    seaweed as, 61-63
    slow-release, 53
        source for, 56
        types of, 56-57
    for trees, 58-60
    types of, 53-55
Frei, Jonathan, on clover as ground
    cover and compost, 44
Fungi, beneficial, 45-47

G

Garden design:
    mulch aid for, 68-70
    no-till, 74-75
Goulin, Francis R., on slow-release
    fertilizers, 56-57
Grass, no-bag mowing of, 89-90
Grass clippings, as fertilizer, 9, 89-90
Green manure. *See* Manure.
Gypsum, as soil additive, 86

H

Hay:
    as fertilizer, 8-9
    nutrient content of, 9
Hill, Stuart, on soil life, 13-15
Hole, Francis D., on soil colors, 10-12
Hydroponics:
    for houseplants, 91-93
    society for, 92
    supplies, sources for, 92

K

Kane, Mark:
    on composting leaves, 29-31
    on soil amendments, 48-51
    on soil-test kits, 16-20
Knoop, William E., on no-bag lawn
    maintenance, 89-90

L

Langsner, Louise, on cover crops for
    building soil, 40-44
Lawns:
    no-bag maintenance of, 89-90
    slow-release fertilizer for, 57, 90
Leaves, as mulch, 67
Lime, as soil amendment, 48-50

M

Manure:
    applying, 9
    green, growing, 85-87
    nutrient content of, 9
Mastalerz, John W., on mulch, 64-67
Mites, as beneficial soil organisms, 14-15
Mowing, no-bag strategies for, 89-90
Mulch:
    applying, 67
    benefits of, 65-66
    drawbacks of, 66
    leaf compost as, 31
    materials for, 66-67, 69
    with newspaper, 68-70
    planting through, 75
    seaweed as, 61-63
    stone, 84
    system for, 74-75
    vertical, 84
    See also Compost.
Mycorrhizae, vesicular-arbuscular, 45-47

N

Nematodes, as beneficial soil
  organisms, 14
Newspaper, as mulch, 68-70
Nutrients, for plants:
  availability of, influenced by pH, 48-50
  concentration in tree foliage, 59

O

Organic residues:
  nutrient content of, 8-9

P

Parnes, Robert, on soil building, 8-9
Paths, newspaper/mulch creation of, 69
Peat moss:
  benefits of, 72
  described, 72
  drawbacks of, 72-73
  environmental issues of use, 73
  as soilless-mix ingredient, 24
  use of in garden, proper, 72-73
Perlite, as soilless-mix ingredient, 24
Perry, Thomas O., on tree roots and
  gardening, 76-79
pH:
  altering with soil amendments, 50
  effect on availability of nutrients, 48-50
  measuring, 49
Pine needles, as mulch, 67
Potting soils:
  buffering capacity of, improving, 25
  compactibility of, 22
  evaluating, 21-25
  nutrients in, retention of, 25
  particle, size of, 22
  peat moss in, 71, 72
  pH of, 25
  porosity of, measuring, 21-22
  soilless, 24
  Styrofoam in, disadvantages of, 22, 24
Propagation, for hydroponics, 92
Protozoa, as beneficial soil organisms, 14

R

Raised beds. *See* Beds.
Reich, Lee, on no-till beds, 74-75

S

Sand, as soilless mix ingredient, 24
Saraceno, Reginaldo, on composting, 39
Sawdust, as soil conditioner, 75
Schultz, Warren, on peat moss, 71-73
Schwab, Suzanne M., on mycorrhizae,
  45-47
Seaweed:
  as fertilizer and soil amendment, 61-63
  products, supply sources for, 62
Seeds, inoculating, 44
Shrubs:
  for moist soils, 27
  planting, with peat moss, 72
Smittle, Delilah, on seaweed soil
  amendments, 61-63
Soil amendments:
  adjusting soil pH with, 48-51
  applying, 51
  improving soil texture with, 50-51
  lime as, 50
  organic matter as, 50-51
  peat moss as, 71-73
  seaweed as, 61-63
  sulfur as, 50
  See also Compost.
Soil organisms:
  encouraging, 15
  role of, 13-15
Soil:
  analyses of, reading, 87-88
  biodynamic preparations for, 85-86
  building, 8-9, 85-87
    with cover crops, 40-44
  colors of, discussed, 10-12
  drainage in, 26-28
  improving, with peat moss, 71-73
  kinds of, discussed, 11-12
  organisms in, 13-15
  pH of,
    adjusting, 50
    testing, 19
  testing, 53, 58
    kits for, 16-20
  texture of, improving, 50
  See also Compost. Drainage. Manure.
    Potting soils.
Soilless mixes, evaluating, 21-25
Soilless system, for houseplants, 91-93
Soleri, Daniela, and Cleveland, David A.,
  on arid-land gardening, 80-84
Springtails, as beneficial soil organisms, 14
Sternberg, Guy, on fertilizing trees, 58-60
Sulfur, as soil amendment, 48-50
Sunken beds. *See* Beds.
Sunshades, for arid-land gardens, 84

T

Thatch, discussed, 90
Tillers, small, for compacted soils, 87
Trees:
  fertilizing, 58-60
    choosing fertilizer for, 58-59
    determining need for, 58
    methods of, 59-60
    nutrients, concentration in foliage, 59
    reasons for, 58
    timing of application, 60
  herbicides on, safe use of, 79
  for moist soils, 27
  mulching, 79
  planting, with peat moss, 72
  roots of,
    discussed, 77-78
    gardening around, 78-79
Trela, Mark, on soil improvement, 85-87

V

VAM (vesicular-arbuscular mycorrhizae).
  See Mycorrhizae.
Vermiculite, as soilless-mix ingredient, 24

W

Water, conserving, garden for, 80-84
Weeding, avoiding, system for, 74-75
Windscreens, for arid-land gardens,
  80, 84
Wood chips, as mulch, 67

The 25 articles in this book originally appeared in *Fine Gardening* magazine. The date of first publication, issue number and page numbers for each article are given below.

8	Organic Matters	September 1989 (9:34-35)
10	The Colorful Soil	November 1988 (4:28-30)
13	The World Under Our Feet	September 1992 (27:50-52)
16	Home Soil Testing	May 1989 (7:54-58)
21	Evaluating Potting Soil	March 1990 (12:22-26)
26	Soil Drainage	July 1993 (32:42-44)
29	Composting Leaves	November 1990 (16:46-48)
32	Get Started in Composting	September 1994 (39:52-55)
36	Making Hot Compost	September 1990 (15:34-37)
40	Cover Crops	November 1990 (16:36-40)
45	Microscopic Partnership	July 1989 (8:48-50)
48	Soil Amendments	July 1991 (20:50-53)
52	A Buyer's Guide to Fertilizers	January 1992 (23:34-37)
56	Slow-Release Fertilizers	March 1993 (30:58-59)
58	Fertilizing Trees Makes a Difference	March 1991 (18:57-59)
61	Seaweed Comes Ashore	November 1991 (22:31-33)
64	A Mulch Primer	January 1993 (29:42-45)
68	Mulch, Don't Dig	November 1989 (10:26-28)
71	Peat Moss	March 1994 (36:51-53)
74	No-Till Gardening	January 1990 (11:52-53)
76	Gardening amid Tree Roots	July 1992 (26:58-61)
80	Managing Water in Arid Gardens	November 1988 (4:14-18)
85	Reclaiming a Lifeless Soil	July 1988 (2:38-40)
89	No More Lawn-Mower Bag	July 1993 (32:48-49)
91	Houseplant Hydroponics	November 1989 (10:56-59)

If you enjoyed this book, you're going to love our magazine.

A year's subscription to *Fine Gardening* brings you the kind of hands-on information you found in this book, and much more. In issue after issue—six times a year—you'll find articles on nurturing specific plants, landscape design, fundamentals and building structures. Expert gardeners will share their knowledge and techniques with you. They will show you how to apply their knowledge in your own backyard. Filled with detailed illustrations and full-color photographs, *Fine Gardening* will inspire you to create and realize your dream garden!

To subscribe, just fill out one of the attached subscription cards or call us at 1-800-888-8286. And as always, your satisfaction is guaranteed, or we'll give you your money back.

Taunton
BOOKS & VIDEOS
for fellow enthusiasts

Taunton Direct, Inc. 63 S. Main Street, P.O. Box 5507, Newtown, CT 06470-5507

FINE GARDENING

Use this card to subscribe to *Fine Gardening* or to request information about other Taunton magazines, books and videos.

☐ 1 year (6 issues) $28
$34 outside the U.S.

☐ 2 years (12 issues) $48
$56 outside the U.S.

(U.S. funds, please. Canadian residents: GST included)

Name _____
Address _____
City _____
State _____ Zip _____

☐ My payment is enclosed.   ☐ Please bill me.
☐ Please send me information about other Taunton magazines, books and videos.

I'm interested in:
1 ☐ sewing
2 ☐ home building
3 ☐ woodworking
4 ☐ gardening
5 ☐ cooking
6 ☐ other   BFG3

☐ Please do not make my name available to other companies.

FINE GARDENING

Use this card to subscribe to *Fine Gardening* or to request information about other Taunton magazines, books and videos.

☐ 1 year (6 issues) $28
$34 outside the U.S.

☐ 2 years (12 issues) $48
$56 outside the U.S.

(U.S. funds, please. Canadian residents: GST included)

Name _____
Address _____
City _____
State _____ Zip _____

☐ My payment is enclosed.   ☐ Please bill me.
☐ Please send me information about other Taunton magazines, books and videos.

I'm interested in:
1 ☐ sewing
2 ☐ home building
3 ☐ woodworking
4 ☐ gardening
5 ☐ cooking
6 ☐ other   BFG3

☐ Please do not make my name available to other companies.